QE
601
C44513
1984

Chishitsu kozo no
keisei.   English.

Geological
structures

61529

Geological Structures

TEXTS IN EARTH SCIENCES

General Editors
Akiho Miyashiro
Seiya Uyeda
Arata Sugimura
Syun-iti Akimoto
Shinjiro Mizutani
Kazuaki Nakamura
Shohei Banno
Hitoshi Mizutani

Orogeny

Akiho Miyashiro, Keiiti Aki and A. M. Celâl Şengör

Geological Structures

Edited by Takeshi Uemura and Shinjiro Mizutani

Geological Structures

Edited by

Takeshi Uemura
and
Shinjiro Mizutani

John Wiley & Sons

Chichester · New York · Brisbane · Toronto · Singapore

Formation of Geological Structures ed. by Takeshi Uemura and Shinjiro Mizutani.
Copyright © 1979 by Takeshi Uemura and Shinjiro Mizutani
Originally published in Japanese by Iwanami Shoten, Publishers, Tokyo, 1979
The English language edition © 1984 by John Wiley & Sons Limited

Library of Congress Cataloging in Publication Data:
Chishitsu kōzō no keisei. English.
  Geological structures.

  (Texts in earth sciences)
  Translation of: Chishitsu kōzō no keisei.
  Includes index.
  1. Geology, Structural. I. Uemura, Takeshi, 1929–   . II. Mizutani, Shin-
jirō, 1932–   .III. Title. IV. Series.
QE601.C44513   1984   551.8   84–3620

ISBN 0 471 90411 2

British Library Cataloguing in Publication Data:
Geological structures.—(Texts in earth sciences).
  1.   Geology, Structural
  I.   Uemura, Takeshi   II. Mizutani, Shinjiro
  III.   Series
  551.8   QE601
ISBN 0 471 90411 2

Typeset by Photo-graphics, Honiton, Devon, England.
Printed by Page Brothers Ltd, Norwich, England.

Contents

Contributors

Chapter 1 Takeshi Uemura
Chapter 2 Shinjiro Mizutani
             Kenji Yairi
             Tokihiko Matsuda
             Kazuo Hoshino
             Yotaro Seki
             Keiichiro Kanehira
Chapter 3 Takeshi Uemura
             Akira Iwamatsu
Chapter 4 Akira Iwamatsu
Chapter 5 Ikuo Hara
             Toshihiko Shimamoto
Chapter 6 Takeshi Uemura
Chapter 7 Keiji Kojima
Appendix Shinjiro Mizutani
             Takeshi Uemura

Preface

This volume attempts to explain several current topics relating to geological structures, bearing in mind the present state of research and the history of structural geology.

The volume is composed of seven chapters, each of which is complete in itself and may be read independently. Chapter 2 is further divided into nine independent and conclusive sections. Problems relating to primary structures in sedimentary or igneous rocks are not considered, not because they present few problems, but because the emphasis in this book is on secondary structures. We also leave the geometry of deformation and continuum mechanics largely to other works, but have endeavoured to provide a readily understandable elucidation depsite the lack of such information.

In Chapter 1 we discuss the history and constitution of geological structures, their division and description, and also outline research techniques and methods with a glance at related fields. This is meant to provide an introduction to the book as a whole.

In Chapter 2 we take several important and moreover, world-renowned, structures and explain the geological features seen in each area, in order to give some examples of the occurrence of actual geological structures, their application and structural geology. The author of each section has visited the place he is writing about and has himself had an opportunity to survey or observe the structures.

In contrast to this, Chapters 3 to 6 take elementary structures such as joints, faults, cleavage or folds, or simple structures such as boudins and focus on the mechanisms by which they are formed. These are explained with the minimum use of elastic theory, fluid mechanics, buckling theory and rheology necessary for a proper understanding, with the intention of discussing the formation of geological structures whilst introducing the analytical views which constitute the important physical characteristics of modern structural geology.

Chapter 3 discusses planes of discontinuity such as faults and joints which are commonly found in rocks, from the viewpoint of clarifying the mechanical properties of the area at the time of their formation. If looked at in terms of the physical properties of the material, the topics mostly involve elastic deformation or brittle fracturing, but to consider the formative processes of geological structures, is a highly relevant problem, basic to understanding how stress fields change with time and the formation of fractures in response to these changes.

The emphasis of Chapter 4 is on slaty cleavage. Many papers have been written on this subject, particularly in recent years, and it continues to be a major problem in the analysis of deformed structures.

Chapter 5, the longest section in the book, deals with 'folds and folding'. It was not until the early 1960s that theories were properly developed to explain folding mechanisms. Since then progress has been remarkable both theoretically and experimentally in addition to observation and survey of natural folds in the field.

Chapter 6 contains some elementary information about the flow of solids, and cites some examples of geological structures formed as a result of the flow of rock. In particular it focuses on current topics such as from what elements flow phenomena are composed, and what their significance is to the formation of geological structures. This chapter is written very much as a contrast to Chapter 3.

Finally, Chapter 7 looks at joints, faults and other diverse discontinuities as discrete surfaces in rocks from the viewpoint of geotechnics. Structural geology incorporates an extremely wide range of applied fields, such as the study of structures associated with petroleum, natural gas and all kinds of other deposits, and also problems relating to the foundations for various types of building or construction projects. The chapter focuses on discrete surfaces in bedrock as a means of considering the relationship between our understanding of the formative mechanisms and processes for geological structures, and how the resulting formations contribute to human life.

This volume thus consists of three differing parts, i.e. Chapter 2, Chapters 3 to 6 and Chapter 7. Chapters 3 to 6 focus on analytical problems, whereas Chapter 2 deals with synthetic topics. In other words, the former discusses the tectonic mechanisms, whereas the latter considers the local or regional structures. However, in both cases, there is a common aim although the standpoints are different, namely a discussion of an individual topic leads to consideration of structures throughout the globe. In contrast to this, Chapter 7 introduces geotechnical topics as an application of structural geology. In this sense, the book is not written in a wholly unified and systematic way. The objective of the editors was rather to indicate the diversity of problems associated with geological structures by introducing the three series of topics, and thereby to make the reader aware of the range of the subject. We hope that the reader will have an opportunity to reconsider the geological structures using the information discussed in each chapter, and thereby that the study of geological structures will be greatly developed.

The study of geological structures began with observing and describing outcrops in the field. To describe complex structures and discuss their formative processes much specialized terminology is used. However, the systematic description and classification of structures are excluded because they are not appropriate to the aim of this volume, instead in the references and further reading given at the end of Chapter 1, the commonest terms are explained and are also demonstrated diagrammatically in the appendix at the end of the volume.

As there are a great many references, only the important ones as indicated by asterisks have been listed at the end of each Chapter.

Takeshi Uemura
Shinjiro Mizutani

Geological Structures
Edited by T. Uemura and S. Mizutani
©1984 John Wiley & Sons Ltd.

1

Introduction

The Austrian geologist Eduard Suess (1831–1914) was the author of a major three-part, four volume book *Das Antlitz der Erde* (The features of the earth) (1885–1909*), whose opening lines declared 'If only an observer could come down from the sky to the earth, and push aside the pink tinged clouds and look around at the earth below.' He would first notice the V-shaped continents stretching southwards and realize that they were reflections of structures in the earth's crust. Nowadays satellite images enable us to see with our own eyes what Suess imagined. Setting aside the distribution and arrangement of large scale land topography such as continents, mountain ranges, great plains, rift valleys, islands and islands arcs, and so on, if we could look down through the oceans at the enormous marine topography underneath, we would see huge mountain ranges encircling the earth as well as series of reliefs extending from the continental margins across the ocean trenches to the level ocean floor. Such things were unquestionably beyond the imagination of Suess and his contemporaries.

The human effort involved in developing our understanding of the organization of such enormous formations little by little, has been incalculable, beginning long before Suess and continuing right up to the present day. As a result we have learned to appreciate something of the intimate relationships that exist between the internal structure of the earth's crust, from which the majority of the large earth features are derived, and the more concrete aspects of structures of rock.

It is quite natural that man, whether now or in ages past, should turn and look at the earth about him and wonder 'How were the mountains formed?' Once a mountain is formed, however, its peaks are gradually worn away over time by external forces of erosion, which act on the earth's surface, and the mountain inevitably ceases to be a mountain. In this respect, structures formed long ago cannot be represented by the present topographic features. In fact all the mountain ranges we see on the earth now, were formed by earth movements in more recent eras, whereas those formed in the Precambrian or Palaeozoic eras have lost their mountainous forms completely. Despite this we are able to establish the fomer presence of mountains with some certainty from the types of strata and structures which are still visible today.

1

Thus, the simple question 'How were the mountains formed?' gives rise to another, 'What is a mountain like inside?', and hence to the birth of orogenic theory. This was developed in the mid-nineteenth century as a result of studies centred mainly on the European Alps, and no doubt was the background which led to the establishment of structural geology as an academic discipline focusing on the structures of the earth.

1.1   The origins of geological structures

Let us first of all consider what *geological structures* are. Any structure seen in the earth's crust, irrespective of its size or scale, is called a geological structure. However, huge structures are clearly also related to the mantle beneath the crust, and so geological structures cannot be limited to those in the crust alone.

Most textbooks and treatises on *structural geology* are visually pleasing to the reader as they use beautiful photographs or illustrations to describe the many diverse 'structures' formed at various times from the Precambrian up to the present day. These usually begin by considering the grandiose folds and faults in the Alps or Rockies and at the same time awaken an interest in the tricks of Nature which have created such formations. As suggested by these photographs and illustrations, in general terms 'geological structures' mean those formed by the distortion and fracturing of rocks as a result of the transmission of the many types of earth movements through the crust. Structures formed in this way are in fact known as '*secondary structures*' in contrast to '*primary structures*' which are formed at virtually the same time as the rocks themselves.

Examples of the latter would be structures seen in sedimentary strata immediately after deposition, which have been completely unaffected by subsequent movements, or those formed when magma solidifies on or beneath the earth's surface. The majority of primary structures thus tend to be formed in a short time under conditions of low solidification. In general however, the term 'geological structures' is used to indicate secondary structures and many textbooks use it exclusively with that meaning.

Let us, for the moment, ignore the distinction between primary and secondary structures and consider folding as a means of examining in more detail the problem of exactly what geological structures are.

Characteristics relating to the physical form of an object are generally divided into its external appearance or features, and internal features. The latter is known as its structure or 'fabric'. Thus geological structure is the structure of the interior of a geological body, and a fold is a fold because of its characteristic internal structure. Size is immaterial in these cases. Whether enormous or minute they are equally folded structures because of their internal configuration. On the other hand, the external appearance of a geological body is closely related to its internal structure yet is subject to considerable change due to external forces of erosion. This is apparent just by looking at the many diverse land forms which make up the external appearance of the crust. It is by no means always a direct reflection of the inner structure.

Let us next consider what elements comprise the internal structure of a geological body. Recognition of the internal structure of a physical object is usually dependent on marks indicating some particular characteristic, and most of these do no more than reflect the arrangement of the grains of material. If the arrangement of the material is uniform in two dimensions then it forms a 'planar fabric', or if it is uniform in one dimension only a 'linear fabric'. Layered structures such as bedding or schistosity are representative examples of geological planar fabrics, but there are other planar structures such as fracture planes in faults or joints, and axial planes of folds which are not dependent directly on the arrangement of the materials themselves. Linear structures may include not only the arrangement of minerals or direction of grain elongation, but also fold axes or the intersection of two planes. At all events all types of structure are composed of planes, lines and combinations thereof as geometric elements, and in physical terms are expressed as 'anisotropy'. In structural petrology, or petro-fabrics, the fabric of rocks is regarded as combinations of such planes and lines, and is classified into four types on the basis of its symmetry (axial symmetry, orthorhombic symmetry, monoclinic symmetry and triclinic symmetry).

Even in the huge and complicated structures seen in the cross-sections of long ranges of folded mountains, the microscopic structures of the minerals are composed of all the various conditions of planes, lines and combinations thereof in just the same way.

1.2 Division of geological structures

There are many types of geological structure. We leave it to other textbooks to describe them systematically, and instead concentrate on two problems thought to be important when classifying extremely diverse, naturally occurring, geological structures and consider their interrelationships.

The first problem is to try to divide the wide range of structures from the simplest to the most complicated, into *elementary* and *compound structures*. Many 'elementary' structures are relatively simple. They contribute to the formation of more complex structures and can be described in terms of planes and lines but there are of course instances when they appear quite independently. Examples of these would be all types of fractured structures such as joints or faults, all types of *foliation* such as rock cleavage or schistosity, and all types of lineation such as the linear arrangement of minerals, fold axes, elongated grains or the intersection of two planes. Folds are formed from both planes and lines, but are usually treated as elementary structures because they are such independent units and are consistent in nature. In contrast, compound structures consist of combinations of elementary structures. For example, structures which are principally folds but with the addition of elements pertaining to faulting may result in *nappes* or *klippen* (rootless blocks), whereas many types of block structures such as *schuppen*, *graben*, *horsts*, tilted blocks and faulted basins may be included in structures which are mostly formed by faulting. Compound structures may arise as a result of two or more types of structure being formed at different times in the same place, or may be composites of differing types of structure formed successively at the initial and final stages of a series of earth movements. Nevertheless, combinations of elementary structures have set patterns, and completely unlimited and random combinations do not occur in nature. We can see demonstrations of these natural laws in the structural arrangement in orogenic belts, and in so-called '*Alpino-type*' structures which are dominated by folding, or in so-called '*Germano-type*' structures which are dominated by faults (Figure 1.1). The same problem is in fact encountered in elementary structures. There are structures which are limited to specific regions and others which appear only at distinct stages in earth movements. Such

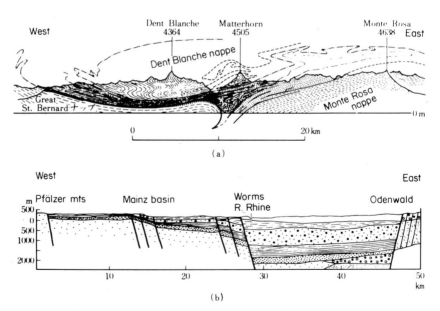

Figure 1.1 (a) Cross-section of Swiss Alps (Pennine belt) showing Alpino-type structure (Heim, 1922). (b) Cross-section of Rhine graben showing Germano-type structure (Billings, 1972)

temporal and spatial characteristics apparent in the distribution of geological structures are direct reflections of the natural conditions of the period and region under which they were formed. They also enable us to clarify changes in time and space of a structure-forming environment.

The second problem is concerned with the *hierarchy of geological structures*, and includes aspects related to their scale, order or unit. The natural world is composed of an infinite number of structural units, but even whilst being organized under their own particular rules, they are yet closely interrelated to form a unified total picture. Nothing in nature is an exception to this, from the largest systems, such as the Galaxy, solar system or the Earth, to the smallest molecule, atom or elementary particle. A number of units making up such a hierarchy are similarly encountered in geological structures. Broadly speaking the largest structures found on Earth are considered to be the layered structure of the crust, mantle and core. Looking solely at the crust, which is of most relevance to geological structures, two major elements may be identified, i.e. the continental and oceanic crusts. The continents may further be divided into stable land masses (cratons), in which disturbances came to an end in much earlier eras, and mountain-forming (orogenic) belts, each of which possesses its own characteristic structures. The continental crust also contains belts of great rift valleys (graben) which are huge fissures of the crust such as seen in East Africa. These form their own particular structural elements. On the other hand, the structure of the oceanic crust is relatively unknown in comparison with the continental crust, but can initially be classified into three major elements comprising the vast ocean floor and long, narrow trenches, to which must be added the mid-oceanic ridges whose peculiarities have been recently clarified. Particularly characteristic structures known as island arcs are found on the boundaries between continent and ocean, the majority of these being concentrated in the West Pacific. They form another unit of structures different from the continents and oceans. These large units may themselves be divided into several structural provinces, each of which may be further subdivided into units of the next scale down, until a single compound structure is finally split into several elementary structural units. Such a *hierarchy* may also be observed amongst individual elementary structures. A detailed examination of a single fault will frequently reveal it to be composed of an *en-échelon* arrangement of several smaller scale faults (Figure 1.2). The repetition of short wavelength synclines and anticlines forming successively large scale *synclinoriums* and *anticlinoriums* are frequently seen in areas of complex folding (Figure 1.3). In other words, the hierarchical position of the structure under consideration must always be borne in mind. It is obvious that the means of recognition in such cases will vary with the scale of the structure. It is common to take the mesoscopic scale (visible with the naked eye) as a standard, and then classify smaller structures as being of microscopic scale, and larger ones as of macroscopic scale.

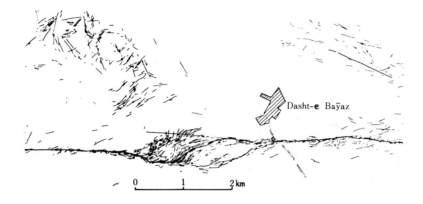

Figure 1.2   Earthquake faults in Dasht-e Baȳaz region (Iran) showing *en échelon* arrangement of numerous small scale faults (Tchalenko and Ambraseys, 1970)

Another important problem in the understanding of structural hierarchy is that of the relationships between each unit, and in particular mutual transfer. For example, it may happen that an orogenic belt is changed into a part of a stable land mass, or an extremely active belt can emerge in the once stable land mass. There are also instances where nappes are formed with the development of recumbent folds, the piles of which give rise to complex folded ranges of mountains such as the Alps. An integration of minute displacements along slaty cleavage may lead to shear folding. Thus relationships in which geological structures mutually transfer in the same rank

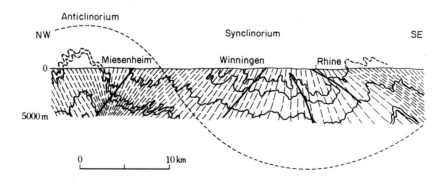

Figure 1.3   Rhineland synclinorium and anticlinorium (Quiring, 1939*)

unit or between differing layers are frequently encountered. This fact is particularly important when considering the processes by which each structural unit was formed.

1.3 Structural geology

Structural geology is the academic discipline for the study of geological structures. Originally the studies of structures and that of composition were allied to give two bases upon which objects could be identified. Looking at this in terms of the earth's crust, the former is structural geology, and the latter corresponds to petrology in its widest sense. The study of developmental processes in the crust, based on the synthesis of knowledge of both structure and composition should be more properly termed 'history of crustal development'.

In English there are two terms which cover this field, the one being 'structural geology' and the other 'tectonic geology' or 'geotechnics'. These are used with almost the same meaning but slight differences of nuance. The former term deals mainly with the morphology and mechanism of individual geological structures, whereas the latter focuses on large scale structures and considers their origins and the processes by which they were formed. Vaguely speaking they should perhaps be called respectively the study of 'elementary geological structures' and 'mega-structure theory' (or 'theory of crustal disturbance'). There is however no definitive boundary between the two. By establishing the mechanisms whereby individual structures were formed it becomes possible to discuss the origins and development of the major structures in which they are found. Conversely, it is often the case that the conditions leading to the formation of individual structures are suggested as a result of information derived from megastructure theory. It becomes even less necessary to maintain any rigid distinction between the two fields when one is considering the relationships of mutual transfer observed between geological structures of different ranks which were described earlier.

Structural geology does not eliminate the object depending on its size or scale of its subject, but it is not usual to apply the term to the layered structure of the globe, i.e. the crust, mantle and core, or to the structure of minerals. However, the more the origins of geological structures are discussed in terms involving the realms beyond the crust, the gradually wider becomes the sphere of structural geology to include the depths of the earth. Structural geology was also said not to fix its means of investigation, and biostratigraphy was the most important tool used in so-called 'classical' structural geology. However, experimental and theoretical research into fracturing and deformation of rocks has advanced considerably, with the result that our knowledge of these fields has now become of vital importance to structural geology. Such situations have brought structural geology closer to geophysics or rock mechanics, and led to the emergence of a new term in the 1960s, '*tectonophy-sics*'.

The first step in studying structural geology is a thorough grasp of the morphological characteristics of structures. With this as a foundation, research into regional structures which aims to establish the spatial distribution and regional characteristics of each type of structure may progress, and on the other hand, the effort may be devoted to understanding the mechanisms by which the structures formed. Synthesizing these clarifies the formative processes and temporal distribution of structures, and also the principles which enable us to picture the structural history.

It was mentioned above that structural geology developed initially with the study of biostratigraphy, but that the modern science makes extensive use of other diverse methods. In the case, for example, of a single structural analysis, it has become possible not only to give simple qualitative assessments but also quantitative treatments, as a result of progress in strain analysis or the mechanical analysis of deformation or fracturing phenomena based on our knowledge of the mechanical properties of rocks. Progress in the analysis of stress–strain relationships inside structural bodies, based particularly on applications of the finite element method, has been quite remarkable in this field. However, the most important feature of modern structural geology must probably be the introduction of experimental techniques.

Experiments in the widest sense may be divided into two quite distinct types. In the first, the behaviour and responses of an object or substance are examined under artificially controlled conditions so as to clarify its properties. Various equipment and apparatus are used to increase the accuracy of the tests. An important experiment of this type in structural geology is the determination of the mechanical properties of rocks or strata using high pressure triaxial testing apparatus. These should perhaps be called '*tests*' rather than 'experiments'. The other type of experiment mainly attempts to clarify the underlying rules of phenomena. It is used in research into the mechanisms and processes by which geological structures are formed, and the majority are so-called '*model experiments*'. They are typically models constructed so as to verify a hypothesis induced from the natural phenomenon. Some models use the appropriate materials and substances, whereas others express the properties or strength, etc. of the object in numerical values. The former type is generally a *scale* model in which the spatial and temporal scale of the natural phenomenon are reduced. For quantitative discussions a relationship required by '*similarity*' must exist between the physical values of both the model and its corresponding prototype. This ratio between them is known as the '*model ratio*'. In cases of slow plastic deformation which is common to many formative processes of geologic structures, the following relationship is obtained from the similarity.

$$C_\eta = C_p \cdot C_1 \cdot C_t$$

C_η, C_p, C_1 and C_t are respectively the model ratios for the values for the equivalent viscosity, density, size and required time (Hubbard, 1937*). Because the size of the model and duration of the experiment are limited by laboratory conditions, materials and substances with appropriate densities and equivalent viscosity are also considerably restricted. In qualitative models the selection of materials is fairly free but even they are restricted to a certain extent. Model substances used in plastic deformation are often various types of clay. The folded structure model of the Appalachian Mountains set up by Willis (1893) became famous but did not take into account the similarity examination. Other experiments based on photo-elastic models utilize the fact that applying a force to a transparent elastic body produces birefringence with phase differences in proportion to the differential stress produced (Brewster's law). Gelatine and epoxy resins are widely used for this.

Numerical models are not restricted as to materials or size, and methods of *simulation* have therefore made considerable progress particularly with the recent developments in high speed computers. At all events, when putting an experiment on a certain phenomenon into practice, whether with a physical or mathematical model, it is essential to construct a system as a model of the phenomenon, namely systems construction. Such systems are composed only of the essential elements left after eliminating all secondary factors from the diverse elements relating to the phenomenon concerned. However, a prerequisite of such decisions is to examine the organization of the phenomenon, namely systems analysis. The conclusions reached as a result of this will lead to a working hypothesis and the construction of a primary model. Once a model is constructed in this way it becomes the departure point for deductive reasoning and consequently, whether or not the natural phenomenon (original) can be explained thereby, will prove the validity or otherwise of the model (hypothesis). If it is found to be invalid then the model must be revised. In other words, model experiments do no more than extract an essential picture of a phenomenon by summarizing all the available knowledge, but they also deepen our recognition of that phenomenon with information deduced from the resulting model.

1.4 Types of crustal movements and tectogenesis

Half a century has passed since the general idea of orogenesis was developed from observations of the features of mountain ranges on lands, or geological structures and rocks seen in parts of the crust which once were mountain ranges. According to Stille (1924*), who gave a clear definition of orogenesis, they occur in comparatively short violent events after long quiet periods and are restricted to fairly limited and narrow zones. Stille recognized that structural discordances, i.e. dino-unconformities, would form even in the process of sequential orogenic movements between the continuously super-posed groups of strata as a result of rapid fold development. He called such a period of violent movement the *'folding phase'* (*orogenic phase*). On the other hand, there is the general idea of *epeirogenesis* which is another type of crustal movement differing from orogenesis. This is a redefinition by Stille (1924*) of the term used long ago by G. K. Gilbert (1890). Its feature is gradual movements over a wide area for a long period of time. Although orogenic movements do form folds and faults, etc., epeirogenesis is largely unaccompanied by such geological structures. In addition to these two types of crustal movement, a third type was proposed by von Bubnoff (1938*). This is known as *diktyogenesis* and is characterized by temporal and spatial scales intermediate between the other two. In particular it relates to the formation of elevated and depressed zones which gives rise to alternating hill and basin topography.

Many different theories have been propounded as to what sort of factors lie behind such differing types of crustal movements, but they may broadly be classified into two series. The first theory is the idea that the crustal movements are derived from forces which radiate out from the centre of the earth, i.e. perpendicularly, producing movements of materials in the same direction. This would mean that a geological structure seen in the upper part of the crust would have originated at depth vertically below. Hence this theory does not recognize large scale horizontal movements of the crust. As a result the proponents of this idea have been nicknamed *'verticalists'* or *'fixists'*. The oscillation or undulation theories of Haarman (1930*), Willis (1929*), and van Bemmelen (1932*, 1935*) are well known and imagine 'gravity-sliding' as the mechanism of the formation of fold belts. The theory of Beloussov (1954*), which also basically considered large scale upward movements of the crust, may also be thought to belong to this series because of its postulation of vertical movement. These ideas were named the 'deep layer tectonics' and theories such as that of Beloussov came to be known parti-cularly as *'block tectonics'*. Although belonging fundamentally to this series of ideas, the expanding earth theory and Ramberg's (1967*) theory are also compatible with the idea of large scale horizontal movements of the crust. The latter theory postulates zones of inverted density in certain horizons under the surface, from which light substances will begin to rise under their own buoyancy and the differences of hydrostatic pressure. When they reach the surface they move horizontally to cause major horizontal displacement of the crust as well as to form Alpino-type structures. It should perhaps be called *'diapir* tectonics'.

By contrast, the other series of ideas looks for the direct origins of geological structures in the relatively shallow parts of the earth and are called the 'shallow layer tectonics'. As they are compatible with the idea of large scale horizontal movements of the crust, its advocates are sometimes called '*horizontalists*' or '*mobilists*'. The historically famous contraction theory proposed by de Beaumont (1852*) and Dana (1847*), and the convection current theories of Holmes (1928*), Vening Meinesz (1933*), Griggs (1939*) and others are all representative examples, and several other theories are derived from these. Other members of this series of tectogenesis may be found, such as the present leader in the field, *plate tectonics*, following the propositions of Isacks *et al.* (1968*) or its prerequisite, the ocean floor spreading theory (Dietz, 1961*). It is interesting to note that all these, with the exception of the contraction theory, have been derived from studies of the oceanic crust and its underlying mantle.

One may say that the modern theories of tectogenesis plainly go far beyond trying to answer such questions as 'How were the mountains formed?' and are not only restricted simply to clarifying the formation of geological structures, but also rapidly deepening our appreciation of the interrelationships between the phenomena encountered in almost all branches of the earth sciences to explain the total phenomenon of the earth.

Takeshi Uemura

References

Beloussov, V. V. (1954): *Osnovnye Voprosy Geotektoniki*, Gosteoltekhizdat, Moskva. English translations available.

Billings, M. P. (1972): *Structural Geology*, 3rd ed., 606 pp., Prentice-Hall, Englewood Cliffs, New Jersey.

Dana, J. (1847): Geological results of the earth's contraction in consequence of cooling, *Am. J. Sci.*, **2**, 176–188.

de Beaumont, E. (1852): *Notice sur les systèmes de montagnes*, 3 tom., P. Bertrand, Paris.

Dietz, R. S. (1961): Continental and ocean basin evolution by spreading of the sea floor, *Nature*, **190**, 854–7.

Griggs, D. (1939): A theory of mountain building, *Am. J. Sci.*, **237**, 611–650.

Haarman, E. (1930): *Die Oscillationstheorie*, 260 S., Ferdinand Enke, Stuttgart.

Heim, Alb. (1922): *Geologie der Schweiz, 11, Die Schweizer Alpen*, 1018 S., Tauchniz, Leipzig.

Hubbard, M. K. (1937): Theory of scale models as applied to the study of geologic structures, *Geol. Soc. Am. Bull.*, **48**, 1459–1520.

Holmes, A. (1928): Radioactivity and earth movements, *Trans. Geol. Soc. Glasgow*, **18**, 559–606.

Isacks, B., Oliver, J. and Sykes, L. R. (1968): Seismology and new global tectonics, *J. Geophys. Res.*, **73**, 5855–99.

Quiring, H. (1939): Uberkontravergente Transformation von Faltenzonen in Rheinischen Gebirge, *Zeitschr. deutsch. geol. Gesellsch.*, **91**, 421–32.

Ramberg, H. (1967): *Cavity, Deformation and the Earth's Crust*, 214 pp., Academic Press, London and New York.

Stille, H. (1924): *Grunfragen der vergleichenden Tektonik*, 443 S., Bertraeger, Berlin.

Suess, E. (1885–1909): *Das Antlitz der Erde*, B. 1778 S., II 703 S., III–I 508 S., III–2 789 S., Tempsky, Prag u. Wien, Freytag, Leipzig.
Tchalenko, J. S. and Ambraseys, N. N. (1970): Structural analysis of the Dasht-e Baȳaz (Iran) earthquake fractures, *Geol. Soc. Am. Bull.*, **81**, 41–60.
van Bemmelen, R. W. (1932): De Undatie theorie, hare afleiding en toepassing op het westelijk deel van de Soenda boog, *Nat. Tijdschr. v. Ned. Indie*, **92**, 85–242.
van Bemmelen, R.W. (1935): The undation theory on the development of the earth's crust, *Proc. 16th Int. Geol. Congr.*, **2**, 965–82.
Vening Meinesz, F. A. (1933): The mechanism of mountain formation in geosynclinal belts, Koninkl. Akad. Wetenschap. Amsterdam, *Proc. sec. Sci.*, **36**, 372–7.
von Bubnoff, S. (1938): Über Gerüstbildung der Erdrinde (Diktyogenese), *Naturwiss.*, **26**, 745–55.
Willis (1929): Continental genesis, *Bull geol. Soc. Am.*, **40**, 281–336.

Further Reading

The following are relatively recent textbooks on general structural geology:
Ashgirei, G. D. (1956): *Strukturackh Geologina*, Moskovskogo Universiteta, Moskva. German translation, Ashgirei, G. D. (1963): *Strukturgeologie*, 572 S., VEB Deutscher Verlag der Wissenschaften, Berlin.
Badgley, P. C. (1965): *Structural and Tectonic Principles*, 521 pp., Harper & Row, New York.
Beloussov, V. V. (1954): *Osnovnye Voprosy Geotektoniki*, Gosteoltekhizdat, Moskva, English translation, *Basic Problems of Geotectonics*, 816 pp., McGraw-Hill, New York.
Billings, M. P. (1972): *Structural Geology*, 3rd ed., 606 pp., Prentice-Hall, Englewood Cliffs, New Jersey.
de Sitter, L. U. (1956): *Structural Geology*, 552 pp., McGraw-Hill, New York.
Hills, E. S. (1972): *Elements of Structural Geology*, 2nd ed., 502 pp., Methuen, London.
Hobbs, B. E., Means, W. P. and Williams, P. F. (1976): *An Outline of Structural Geology*, 571 pp. John Wiley, New York.
Metz, K. (1967): *Lehrbuch der tektonishchen Geologie*, 357 S., Ferdinand Enke, Stuttgart.
Schmidt-Thomé, P. (1972): *Jahrbuch der allgemeinen Geologie, Bd. II: Tektonik*, 587 S., Ferdinand Enke, Stuttgart.
Spencer, E. W. (1977): *Introduction to the Structure of the Earth*, 2nd ed. 640 pp., McGraw-Hill, New York.

2

Examples of Geological Structures

The continents of the world are divided fundamentally into stable lands of the Precambrian age and mobile belts of later ages. The Precambrian era was very much longer than all the subsequent ones and accounts for some five-sixths of the history of the earth's crust. Despite this, the Precambrian rocks are not extensively and uniformly distributed over the entire surface of the earth, but are limited to continental regions as continuous basements. These are the so-called '*shields*' (Figure 2.1).

Sedimentary formations deposited from the Palaeozoic onwards are found around such shields. The strata are generally thin, almost horizontal, and little deformed. Such areas are called '*platforms*'. In the shields and platforms, subsidence has occurred, in places forming huge inland basins. These movements were gentle but of very long duration and spread over a wide area.

In each continent, mobile belts subjected to orogenic movement have been developed around the shields and platforms since the beginning of the Palaeozoic era. In these areas the strata are folded in extreme fashions, sometimes highly metamorphosed, and granitic plutonic rocks are commonly found.

Successive changes of tectonic evolution everywhere on earth take place to lay the result of younger movement on that of the older one, and it is by no means certain that an identical structure will ever be formed under absolutely identical conditions again. A comparative study shows however, that similar structures are in fact formed in different places and at differing times. Thus it seems likely that a certain type of structure is formed under a certain specific geological condition. By studying a tectonic history and by identifying general and common features developed in each belt or continental region, one can notice that many different types of geological structures are recognized on the surface of the earth. In this chapter we discuss several types of such geological structures and intend to explain what kind of background factors result in what type of structure.

First of all we discuss faults and associated structures such as the East African Rift System, the San Andreas fault and the block structures of Eastern China. We then turn to the Swiss Alps and the Canadian Rockies in

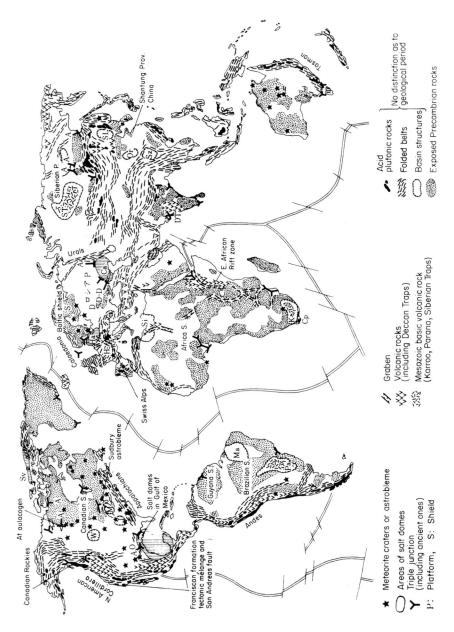

Figure 2.1 (see opposite)

Figure 2.1 The geology and geological structures of the continents.

Rocks of Precambrian age are found over huge areas on every continent, forming shields or platforms. Examples include the *Canadian Shield, Baltic Shield, Russian Platform, Siberian Platform, African Shield, Guyana Shield* and *Brazilian Shield*.

Mobile belts are found surrounding these areas: the Caledonian orogenic belt in northwest Europe, the *East Greenland belt*, and the Appalachian orogenic belt were formed in the early Palaeozoic. The *Hercynian (Variscan) orogenic belt* (H) formed in the late Palaeozoic also runs across Europe. Part of the *Appalachian orogenic belt* was again affected by movements in this period, together with the *Ouachita orogenic belt* (O) in its southwest extension. Strata comprising mainly the South African Cape System (Cp) and its corresponding Silurian-Devonian systems are distributed from the Falkland Islands to southern South America and together with the Tasman orogenic belts of Australia they formed an orogenic belt around the continental margins of Gondwana. It seems likely that the Palaeozoic mobile belt from NovaYaZemuria to the *Urals* was connected with the Mongolian mobile belt in China. Palaeozoic metamorphic rocks are also found in the North American Cordillera and the *Andes*, but we do not understand in detail the movements to which these regions were subjected from the Mesozoic to the Cenozoic.

The *Alpine orogenic belt* (A), formed in Mesozoic to Cenozoic times, runs from the Pyrenees and margins of the Mediterranean, through the Carpathians and via the Zagros Mountains in Iran to join up with the Himalayans, which surround the northern part of the Indian sub-continent.

Huge sedimentary basins have at some time or other been formed in every continent. Typical of these are the *Williston* (W), *Michigan* (M), and *Illinois* (I) *basins* in North America. The *Maranhão (Ma)* basin in South America or the basin which continues into the *Gulf of Sirte* in North Africa (Si) are probably of similar size. In China there are various basins such as the *Sung Liao* (SL), *Ordos (Or)* or Szechuan (Sz).

Both the *Karroo* (K) and *Parana* (P) basins contain large amounts of basic volcanic rocks, of mainly Jurassic age, whilst eruptions in the *Siberian traps* range from Permian to Triassic in age.

The *Hanover (Ha) Basin* is an area well known for the development of salt domes. These have also been reported in the Pyrenees, Transylvania (T), Dnepr (D), the northern Caspian Sea (Ca) and Iran and North Africa. For evaporites and salt domes around the Atlantic margins, see Figures 2.VIII.10 and 2.VIII.11. The intrusion of evaporites into Mesozoic and Palaeozoic strata in the Canadian Arctic Sverdrup basin (Sv) has produced complex fold structures.

In East Africa one finds a graben structure known as the Gregory Rift, and in central Europe, the *Rhine Graben*. Remarkable eruptions of volcanic rocks can be seen along them. Cenozoic volcanic rocks emerged mostly in island arc areas, but they are also distributed in the west of North America and the Andes belt. Eruptions in the Deccan Traps (DT), from the Cretaceous to the Tertiary are similar to volcanic rocks in the Karroo basin described above, and are thought to have been related to the break-up of *Gondwana*. On the other hand, volcanic rocks in Iceland are directly related to activity along the mid-Atlantic Ridge.

Triple junctions with three rift arms including such an ancient example as the Athapuscow (At), or *Dnepr-Donetsk* (D-D) *aulacogens* are based on Burke and Dewey (1973). For the location of *astroblemes*, see Table 2.IX.1.

the *North American Cordillera* as examples of fold and over-thrust zones. It scarcely needs saying that the results of recent developments in plate tectonics have forced us to reappraise our ideas about all the mobile belts, and examine their tectonic development from a global viewpoint. In this reassessment so-called 'tectonic *mélange*' and 'aulacogen' have attracted particular attention. In this chapter we discuss them with reference to the West Coast of California and the Great Slave Lake in Canada, respectively.

Since the geologic body is always within the earth's gravity field, the specific gravity of rock plays an essential role in deformation process, and the physical properties of the rock itself are the direct cause of a structure. A typical one is a dome or fold structure made by the plastic flow of rock salt with a low specific gravity. An often cited example is the area of salt domes of Gulf Coast which is also well known as an oilfield.

Our final topic in this discussion of geological structures deals with a meteorite crater. In several places on the surface of continental regions, geological structures have been found whose form closely resembles that of a volcano, although there are no signs of past volcanic activity. An interpretation that these are the result of impact by meteorites has been supported by the presence of high-pressure minerals generated by shock wave at the time of impact. Similar structures formed in the earlier geological periods offer no such evidence, but some clues are afforded by the structure itself, the types of rock found in it, and their distribution. In this chapter we explain these structures taking the remarkable Sudbury Astrobleme as our example.

In Figure 2.1 we present a summary of the world's geology and geological structures. The examples we actually describe are no more than a very small part of all the representative structures encountered throughout the world, but they will be found helpful for understanding the geological structure and tectonic history of many other areas in the world.

Shinjiro Mizutani

I The East African rift system

A general look at the configuration of the African continent shows it to be composed of several wide, low basins with high plateaus around them. Such a landform is peculiar to continental shield areas and is known as 'basin and swell' structure (Holmes, 1965). The plateau land of eastern Africa is an area where the 'swell' has become particularly pronounced. The *East African rift zone* runs for some 4000 km north to south, from Ethiopia to Mozambique, through the centre of this elevation (Figure 2.I.1). The huge graben seen in East Africa are called 'rift valleys' and Gregory (1896) was the first to

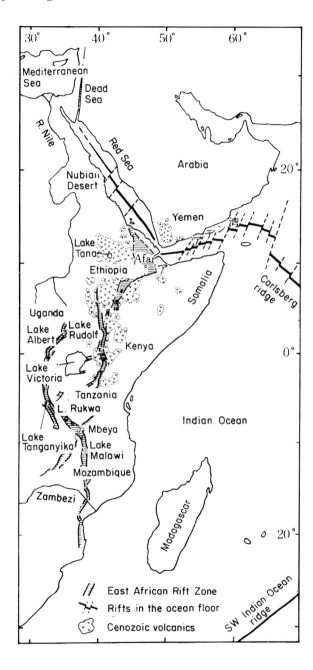

Figure 2.I.1   Afro-Arabian Rift system (East African Rift zones Gulf of Aden – Red Sea – Dead Sea rifts) (after McConnell (1972*) and Pilger and Rosler (1976*))

establish that these are in fact narrow depressions limited by several normal faults. The huge depressions are found not only in the interior of the African continent but also in the Red Sea and Gulf of Aden, which form the boundary between the continent and the Arabian Peninsula. The Red Sea rift is connected to the *Dead Sea rift* to the north, whereas the Gulf of Aden rift extends into the Indian Ocean to the east and is connected to the rift system of the Mid-Indian Ocean ridge via the Carlsberg ridge. This series of rift systems, which runs for more than 6500 km, is known as the *Afro-Arabian rift system*.

The average altitude of the East African plateau is about 1000 m, and in its centre is a huge basin occupied by the shallow waters of Lake Victoria. Both the east and west sides of this basin are bounded by further elevated areas 1000–2000 m higher than the plateau itself. The East African rift zone is divided into two parts, the Eastern Rift and the Western Rift, following these two elevations. The Western Rift is composed of a number of narrow lakes beginning with Lake Albert in the north to Lake Malawi in the south. The areas of rift valley occupied by lakes are arranged *en échelon*. Where the Lakes Albert and Edward approach, Mount Ruwenzori (5109 m), covered with glaciers even though it is near the Equator, soars up to form an enormous horst block composed of Precambrian rocks. Lake Tanganyika is the deepest lake after Lake Baikal in Siberia, its floor being below sea level as is that of Lake Malawi. The height difference between the rift shoulder and the lake floor is as much as 3300 m for Lake Tanganyika and 2600 m for Lake Malawi.

Saggerson and Baker (1965) investigated altitude variations in preserved erosion-surfaces, and established that West Kenya and Northern Tanzania have risen nearly 1800 m since middle Tertiary and that the form of the elevation was an elongated dome in a north–south direction. This uplifted belt is known as the Kenya dome, and a similar area in Ethiopia, is the Ethiopian dome. Following the crest of these two domes, the Eastern Rift zone extends north or north-northeast as far as Afar, whereupon it becomes a fan-shaped depression opening to the north. The Eastern Rift which extends into Kenya and Tanzania is called the Gregory Rift after J. W. Gregory. The *Gregory rift* is indistinct in Tanzania, but runs southwest as an *en échelon* arrangement divided into several depressions. The rift floor is relatively flat, but separated into several basins by small transverse elevations. The majority of the lake basins in the Eastern Rift area are shallow and not the kind of deep lakes encountered in the Western Rift. There are also crater lakes and lakes dammed by lava. Many lakes form isolated internal drainage basins and tend to have high salt concentrations because of the extremely dry climate. At Lake Magadi in southern Kenya the evaporites formed under these conditions are exploited for salt.

The rift valleys are bounded by several normal faults; in the Gregory Rift the main fault scarps are 1500 m in height and form a graben about 50 km wide (Figure 2.1.4). The total throw of the marginal faults is presumed as much as 300 to 4000 m. In the East African rift zone there is a remarkable uniformity in width of the rift valley to be between 40–60 km. Other intracontinental rifts, such as the Rhine *Graben* or Baikal rift, have similar dimensions. The idea that these widths have the same order as the thickness of the continental crust has been presented by analogies with the model experiments of Cloos (1936) and mechanical analysis by Heiskanen and Meinesz (1958).

(a) Outline of geology

The Tanzania Shield, which spreads out around Lake Victoria, is bounded on the east and west by two Precambrian metamorphic zones known as the 'Mozambique metamorphic zone' (835–400 Ma) trending north–south, and the 'Ubendian metamorphic zone' (2100–1950 Ma) trending northwest–southeast (Cahen and Snelling, 1966). The Eastern Rift develops along the western edge of the Mozambique metamorphic zone and the Eastern Rift follows the Ubendian metamorphic zone. Such facts led McConnell (1972*) to emphasize that the location and the trend of Cenozoic systems were considerably influenced by the distribution and structural trends of Precambrian metamorphic zones.

In the area from Ethiopia and Somalia to the Yemen on the southern Arabian peninsula, marine transgression from the Indian Ocean has repeatedly occurred since the Triassic. In the late Eocene, the uplift of the area brought about a sudden regression (Beydoun, 1970). An eruption of alkaline basalts of the fissure eruption type accompanied this uplift and covered the Ethiopia–Yemen region over the Red Sea, to form a huge lava plateau. The plateau basalts known as the Trap Series, are several hundred metres thick on the plateau and exceed 2000 m on the rift margins. In other words, the uplift of the basement and eruption of plateau basalts occurred at about the same time in the early Tertiary and accompanied movement along the central axis of the uplifted zone. This was the first stage of the rift development in Ethiopia (Mohr, 1971*). The activity of shield volcanoes of the central eruption type began in the early Miocene on the Ethiopian plateau and at about the same time there were flows of basalt in the *Afar depression*. In the Pliocene, welded tuff covered central southern Ethiopia. Until this time the *Ethiopian Rift* had been no more than a shallow depression, but with the major uplift of the Ethiopian dome in the early Pleistocene, a graben was formed along the central axis of the dome. At about the time of these movements there was also some basaltic to acidic volcanisms along the rift axis. After the Pleistocene the rift floor was severely broken up and the *Wonji fault belt* was formed along the rift axis. The Wonji fault belt, 3–15 km wide, is characterized by normal fault swarm associated with open tensional fissures and Quarternary caldera volcanoes. The belt tends to be arranged *en échelon* along the rift axis (Figure 2.I.2).

The earliest volcanism in the Gregory Rift began with the Miocene basaltic flows in the Turkana depression in northern Kenya (14–23 Ma). At about the same time there was alkaline volcanic activity of the central eruption type in the Kenya–Uganda border region. An alkaline complex accompanied by

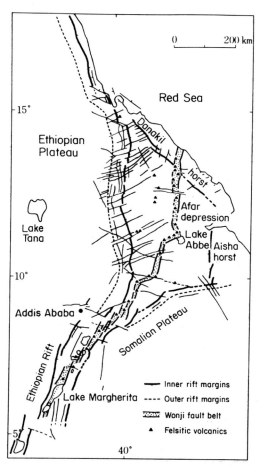

Figure 2.I.2.   Volcanic and tectonic lineaments in the Ethiopian Rift and Afar depression (after Mohr (1971*))

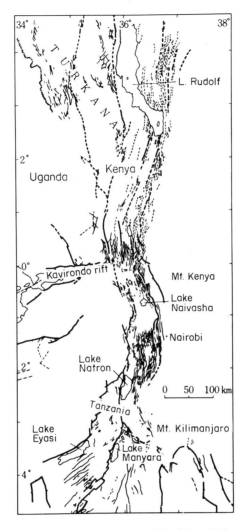

Figure 2.I.3.   Fault pattern in the Gregory Rift (after Baker *et al.* (1972*))

carbonatites is seen in the *Kavirondo Rift* (Figure 2.I.3), which branches off the centre of the Gregory Rift in a westerly direction, and also dates from this period. In the later Miocene (11–14 Ma) there were flows of phonolite of the fissure eruption type, accompanying uplift of the Kenyan dome (about 300 m) and these covered wide areas of southern Kenya. An asymmetrical graben (Figure 2.I.4) bounded by faults only on its west side was formed following the crest of the dome. In the Pliocene, trachyte and basalt erupted on the rift

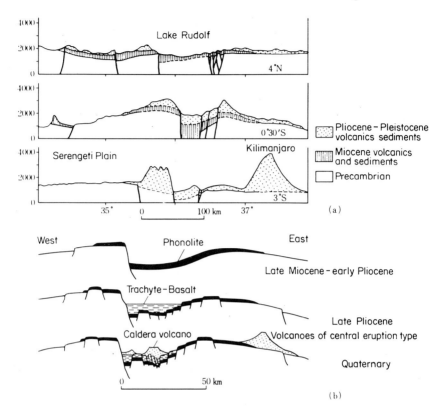

Figure 2.1.4   (a) E–W geological profile of the Gregory Rift. (b) Schematic profile
    showing rift development (after Baker *et al.* (1972*) and Baker (1975))

floor and ignimbrite flowed over the rift shoulders on to the plateaux. At
about this time isolated, huge central volcanoes such as Mount Kenya or
Kilimanjaro began to build up on the eastern flank of the Kenyan dome, and
their activity continued right up until the Pleistocene. By the end of the
Pliocene period the Kenya dome had risen a further 1500 m and the rift
structures along the central axis of the dome were almost complete. Volcanic
activity and faulting in the late Pleistocene seems to have been limited to the
rift floor and resulted in the development of a series of small calderas and a
dense swarm of minor faults at 2–3 km intervals (Baker *et al.*, 1972, Williams,
1970*). In October 1960 Dawson observed carbonatitic lava flowing out from
the Oldoinyo Lengai volcano in the southern Gregory Rift (Dawson, 1962).
This event demonstrated that carbonatite magmas exist in nature, and it was
noticed that the location of the activity coincided with the rift zone. In
northern Tanzania where this volcano erupted, as in the Kavirondo Rift,

many more earthquakes have been recorded than in the other areas in East Africa.

Volcanic activity in the Western Rift is limited to three small areas located between Lake Albert and Lake Tanganyika, and to the Mbeya region positioned at its intersection with the Eastern Rift. All have been active since the Pliocene and the Nyamlagira and Niragongo volcanoes to the north of Lake Kivu are still active today. As the Western Rift shows almost no signs of being covered with volcanic rocks, the Precambrian basement is exposed directly from the rift shoulder to the margins. The oldest sediments in the rift floor are shallow lake or marsh sediments from the early Miocene (Hopwood and Lepersonne, 1953). They suggested that this area had formerly been one of gentle downwarping. By examining drilled cores and as a result of geophysical works carried out in Lake Albert, the maximum thickness of sediments beneath the bottom of the shallow lake was assessed to be 2500 m (Harris *et al.*, 1956). However, the sediments in Lake Malawi and southern Lake Tanganyika are extremely thin (Degens *et al.*, 1971). The major rift-faulting which formed the graben structures began in the Pliocene (Bishop, 1965), and in contrast to the Eastern Rift, there continues to be considerable seismic activity in the northern part of the Western Rift (Wohlenberg, 1969). In the major Toro earthquake (M = 6.7) in western Uganda, an earthquake fault with a throw of 2 m was observed running for a distance of 20 km parallel to the rift. (Loupekine, 1966).

(b)   Deep structure of Rift Valleys

The epicentre of earthquakes (Figure 2.I.5), which occur along the rift zone tend to be mainly within 30 km of the surface (Maasha, 1975), although at the eastern margin of the Ethiopian plateau the epicentres extend down to 60 km (Gouin, 1970). These facts demonstrate that the fracture phenomena pertaining to rift-valley formation occur at relatively shallow positions between the crust and the highest part of the upper mantle.

Figure 2.I.6 shows the topography and Bouguer gravity anomaly along the E–W profile across the east African plateau, and models of lithospheric structure. Apart from local anomalies seen in the Eastern and the Western Rifts, the uplifted area of the east African plateau is characterized by a broad negative Bouguer anomaly of wavelength 1000 km, and maximum amplitude –150 mgal. The small gradient of these regional anomaly curves implies that lighter substances, which give rise to the negative anomaly, are present deep down and the density contrast between these substances and the surrounding rock is also small. On the other hand, seismic studies indicate that the crust in East Africa is 35–40 km thick, as is typically the case in the continental crust (Vp = 6.6 km/s); that there is a low-velocity layer with high attenuation for S-waves beneath the upper mantle layer (Vp = 8.0 km/s); and that the upper

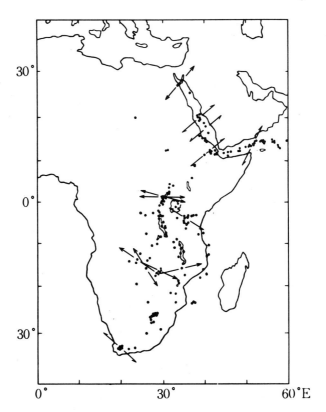

Figure 2.I.5   Seismicity and possible stress fields given by earthquake mechanisms on African continent. Black dots — epicentres 1963–1970. Arrows — direction of crustal extension obtained from earthquake mechanisms (after Fairhead and Girdler (1969*) and Maasha and Mdnar (1972)).

surface of the low-velocity layer rises to a level of 80 km underneath the Eastern Rift. The lithospheric models in Figure 2.I.6 have been constructed on the basis of this data. The model suggests that the lithosphere becomes the thinner under the East African plateau, and particularly so under the Eastern and Western Rifts.

In the Eastern and the Western Rift the regional negative anomaly is superimposed by the local Bouguer negative anomalies with wavelength of 150–200 km, and amplitudes of 50–70 mgal. On the rift axis relatively positive anomalies, 40–80 km wide, with amplitudes of 30–60 mgal are superimposed on the local negative anomalies in the Eastern Rift. The local negative anomalies in the Western Rift were discovered at Lake Albert by Bullard (1936). Girdler (1964) explained them as a combination of subsidence of a crustal block bounded by normal faults, and thick layers of low-density sediments embedded in the graben. The local negative anomaly in the *Gregory Rift*, on the other hand, is thought to be due to thick layers of younger volcanic rocks with low density which cover the shoulders and the floor of the rift. The pronounced positive anomaly along the central axis of the Gregory Rift, and the steep gradient of its curve indicates that material with a positive density contrast relative to the adjacent crust is intruded into a fairly shallow position. The model in Figure 2.I.6(b) suggests that a 10

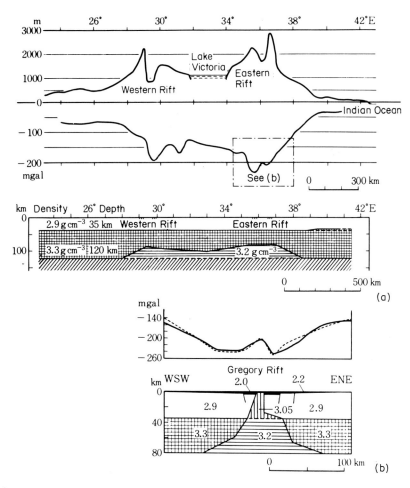

Figure 2.I.6 (a) E–W profile at 0.5 °S across East African plateau, Bouguer anomalies and lithospheric model (after Wohlenberg (1976*)). (b) Bouguer anomalies and crustal model of the Gregory Rift. Figures refer to density (after Baker and Wohlenberg (1971*))

km-wide basic igneous intrusion reaches a depth of 2.5 km below the surface of the rift floor. The apparent crustal dilatation reaches up to 10 km in the model and corresponds well with the crustal extension (5–10 km) deduced by Baker and Wohlenberg (1971*) from the throw of the rift faults near the earth's surface. This relative positive Bouguer anomaly may be traced continuously as far as the central region of the Gregory Rift (Searle, 1970) but disappears in the vicinity of 2° south and cannot be found any further south (Darracot *et al.*, 1972). A similar positive anomaly is found in Ethiopia, following the Wonji fault belt, where the width of the anomaly is wider than that in the Gregory Rift and has been interpreted as basaltic dyke swarm intruding to a shallow level (Makris *et al.*, 1970).

Mohr and Wood (1976*) estimated the thickness of the crust and lithosphere from the spacing of volcanoes distributed along the Eastern Rift and from seismic and gravitational data (Table 2.I.1). They concluded that the thickness of both was less under the rifts than under the plateau, and that the

Table 2.I.1   Volcano spacings, thickness of lithosphere and crust in East African Rift zones (after Mohr and Wood (1976*))

Location	Age of volcano	Spacings of volcanoes (km)	Crustal thickness (km)	Lithospheric thickness (km)
Erta-ali volcanic series (NW edge of Afar depression)	Quaternary	10±3	16	16
Dubbi volcanic series (Northern Afar depression)	Pliocene-Quaternary	19±6	18−25	?25
Inner Ethiopian Rift	Pliocene-Quarternary	43±13	20−35	35−50
Inner Gregory Rift	Pliocene-Quaternary	42±11	20−35	35−50
Western shoulder of Gregory Rift (E. Uganda)	Miocene	72±9	?35	75−80
Western shoulder of Ethiopian Rift	Miocene-Pliocene	70±10	45−50	80
Ethiopian Plateau	Oligocene-Miocene	109±22	30−40	?80−120

thickness under the rifts declined in a northerly direction, i.e. in the order Gregory Rift, *Ethiopian Rift, Afar depression*. The idea that the rift system was formed as a result of a tensional stress field at right angles to its strike has been held by many people since Gregory (1896) in view of the development of normal faults observed at the surface, and volcanic activity of the fissure eruption type. The crustal dilatation or opening due to the intrusion of mantle material or lithospheric attenuation seen throughout the rift systems do not contradict this idea. It is given further support by the fact that fault plane analysis of earthquakes occurring along the rift systems shows normal fault mechanisms (Figure 2.I.5). Recent measurements in the Ethiopian Rift using geodimeters have suggested that the surface extension is taking place at an average rate of 3–6 mm year in a direction at right angles to the strike of the rift (Mohr, 1977).

(c) Break-up of the continental crust and the development of the rift systems

The close association of intracontinental rift systems with alkaline magmatisms is said to be quite a common phenomenon (Le Bas, 1971). The volcanic activity of the East African rift system is also characterized by remarkable alkaline magmatism such as fissure eruption of phonolite or alkaline basalts which formed extensive lava plateaus in Ethiopia and Kenya and also voluminous trachyte and ignimbrite flows and nephelinite-carbonatite volcanism. The further south one goes from Ethiopia into the interior of the continent, the more marked is the tendency of volcanic rocks in the rift valley to be alkaline (Harris, 1969). Furthermore, although basalts in the Red Sea and Gulf of Aden are the tholeiitic basalts typically found in the oceanic crust, in the Afar depression the basalts are transitional between the tholeiitic and the alkaline types (Gass, 1970). The ratio of the volume of basaltic and acidic volcanics is 6:1 in Ethiopia, but 1.3:1 in southern Kenya (Williams, 1972; Mohr, 1968). Such changes in the chemical composition of the volcanics from north to south are fully concordant with the change in the volume of the total volcanics (345 000 km^3 in Ethiopia, 144 000 km^3 in Kenya), the change in the amount of domal uplift of the crust (3000 m in Ethiopian dome; 1800 m in Kenya dome) and the progressive decrease from north to south of the throw of the rift faults. These are also concordant with the changes in crustal dilatation and lithospheric attenuation from north to south. Changes in doming, rift-faulting, alkaline magmatism, lithospheric attenuation, and so on may therefore all be the expressions of the same major process and viewed as a series of phenomena very intimately related in their origins (Gass, 1970).

Pilger and Rosler (1976*) discussed the progressive phases of topographic development in time and space which had contributed to the evolution of the rift systems in the Dead Sea, Red Sea, Gulf of Aden and East Africa (Figure 2.I.7). The Dead Sea and the Red Sea were the first to be affected, followed

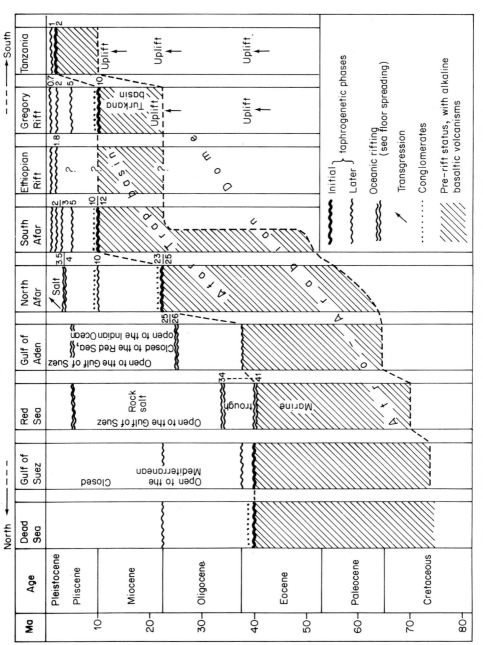

Figure 2.1.7 Development of Afro-Arabian rift system (after Pilger and Rosler (1976*))

by the Gulf of Aden and north Afar, and progressing from south Afar and Ethiopia to Kenya and Tanzania. They attributed such progression or wandering of the rifting to a decrease in the amount of the lithospheric attenuation and crustal fracturing from north to south. They claim that the Red Sea and the Gulf of Aden have already reached rifting stage of oceanic ridge development, and the northern Afar is at the earlier stage in sea floor spreading. According to this idea, the East Africa Rift System may be in the very earliest stage of the development of an intracontinental rift and an oceanic ridge, representing the break up of the continent, and the further south in the system the more embryonic the stage. Earthquakes are the most sensitive indicators of crustal fracturing. The feature of the rift systems as topographical depressions disappears at about 20° south, but the seismic zone extends much further south (Figure 2.I.5). On the basis of these facts Fairhead and Girdler (1969[*]) pointed out that areas of incipient rifting perhaps extends as far as South Africa.

The application of the *plate tectonic* model to studies of the evolution of the Red Sea and Gulf of Aden rifts have made rapid progress since 1960 with the accumulation of wide ranging data from investigations of submarine topography, magnetic anomalies, gravity anomalies, heat flows, seismic exploration, cores from the ocean floors and geology (Le Pichon and Heirtzler, 1968; Vine, 1966; Matthews *et al.*, 1967; McKenzie *et al.*, 1970[*]). According to research, the ocean floors of the Red Sea and the Gulf of Aden were created by a process of sea-floor spreading as seen in mid-oceanic ridges, and the history of this development may be explained systematically by the anticlockwise rotation of the Arabian plate with respect to the African plate. The history of the sea floors of the Red Sea and the Gulf of Aden prior to the Miocene is not so obvious, but Girdler and Styles (1976[*]) used recent drilling data and magnetic anomalies for the Red Sea floor to show that its inception went back 40 Ma. The relative motion of these two plates consistently explains the large lateral displacement along the *Dead Sea Rift* (Quennell, 1958; Freund *et al.*, 1970). It is also said to be useful in interpreting the complex history of the Afar depression where the continental crust encounters the oceanic one (Barberi and Varet, 1975).

According to plate tectonics, the East African Rift System is also an opening plate boundary, the system dividing the African plate into the Nubian plate in the west and the Somalian plate in the east (Gass and Gibson, 1969). In other words the three rift systems of the Red Sea, the Gulf of Aden and East Africa come together at the Afar depression to form a *triple junction* of the RRR type described by McKenzie and Morgan (1969). McKenzie *et al.* (1970[*]) computed the pole of rotation for the East African Rift System to be 8.5°S 30°E at the centre of the Western Rift and an angle of rotation of 1.9° from significant differences in the positions of the spreading centre for the Red Sea and the Gulf of Aden. Their calculation gives crustal extensions across the Eastern Rift of 65 km in the Ethiopian Rift and 30 km in the centre of the Gregory Rift. In contrast, the amount of crustal extensions deduced from geological and gravitational data is 30 km for the *Ethiopian Rift*, 10 km for the *Gregory Rift* and 2–3 km in northern Tanzania (Baker *et al.*, 1972*). To explain the latter values, attempts have also been made to place an East African pole of rotation further south in the rift system (Darracott *et al.*, 1972). These attempts based on kinematics of plate motion may form a useful hyothesis to explain the progressive evolution of rifting in East Africa from north to south. When the values of crustal extensions shown above are converted into average spreading rates, they represent 2 mm/year in Ethiopia and 1 mm/year in Kenya (Wohlenberg, 1976*). These values are one order of magnitude smaller than the spreading rate for the Red Sea and the Gulf of Aden Rifts, and nearly two orders smaller than those of typical mid-oceanic ridges. The phenomenon of lithospheric attenuation is thus of considerable interest when considering such small rates of deformation, as it suggests plastic deformation of the plate in the early stages of the break-up of the continent.

Pilger and Rosler (1976*) identified seven tectonic events in rifting in Afar and the surrounding area, and indicated that the interval between each event became progressively shorter. Certainly, when the length of the quiet periods becomes zero it will mark the start of the sea floor spreading in the true sense of the word and after this the spreading rate would probably accelerate. On the other hand, Burke and Dewey (1973) have postulated the development of a plate boundary or a triple junction of a RRR type in the continental crust based on several examples, saying that three plate boundaries would not always develop uniformly and that one of them would become inactive as a failed rift (*aulacogen*), left behind at a certain stage of the evolution. The super continent of Gondwanaland, with the African continent as one of its components, has experienced such events of great interest to the history of the earth as break-up and drift of the continent. Will the East African rift system indeed give birth to a new ocean and a Somalian continent separate from the Nubian one and drift into the Indian Ocean? (Dietz and Holden, 1970).

Kenji Yairi

References

Baker, B. H. and Wohlenberg, J. (1971): Structure and evolution of the Kenya Rift Valley, *Nature*, **229**, 538–42.

Baker, B. H., Mohr, P. A. and Williams, L. A. J. (1972): Geology of the eastern rift system of Africa, *Geol. Soc. Am. Spec. Paper*, 136, 67 pp.

Fairhead, J. D and Girdler, R. W. (1969): How far does the rift system extend through Africa? *Nature*, **221**, 1018–20.

Girdler, R. W. and Styles, P. (1976): The relevance of magnetic anomalies over the southern Red Sea and Gulf of Aden to Afar, *Afar between Continental and Oceanic Rifting*, A. Pilger and A. Rosler (eds), Schweizerbart., Stuttgart, 156¿70.

McConnell, R. B. (1972): Geological development of the rift system of eastern Africa, *Geol. Soc. Am. Bull.* **83**, 2549–72.

McKenzie, D. P., Davies, D. and Molnar, P. (1970): Plate tectonics of the Red Sea and East Africa, *Nature*, **226**, 243–8.

Mohr, P. A. (1971): Outline tectonics of Ethiopia, *Tectonics of Africa*, Unesco, Paris, 447–58.

Mohr, P. A. and Wood, C. A. (1976): Volcano spacings and lithospheric attenuation in the Eastern Rift of Africa, *Earth Planet. Sci. Lett.*, **33**, 126–44.

Pilger, A. and Rosler, A. (1976): Temporal relationships in the tectonic evolution of the Afar Depression (Ethiopia) and the adjacent Afro-Arabian rift systems, *Afar between Continental and Oceanic Rifting*, A. Pilger and A. Rosler (eds), Schweizerbart., Stuttgart, 1–25.

Williams, L. A. J. (1970): The volcanics of the Gregory rift valley, East Africa, *Bull. Volcanol.*, **34**, 439–65.

Wohlenberg, J. (1976): The structure of the lithosphere beneath the East African Rift Zones from interpretation of Bouguer Anomalies, *Afar between Continental and Oceanic Rifting*, A. Pilger and A. Rosler (eds). Schweizerbart, Stuttgart, 125–30.

II   *The San Andreas fault*

The Great Californian Strike-slip fault

The existence of the *San Andreas fault* began to attract the attention of Californian geologists in the 1890s. The name 'San Andreas fault' is said to have been introduced in 1895 (in a paper by Lawson; Crowell, 1962*). This was about ten years later than the discovery and naming of the Median Tectonic Line in Japan.

It was the San Francisco earthquake of 1906 that overnight brought fame to the fault. Displacements up to a maximum of about 7 m appeared along a distance of approximately 430 km, following the line of the fault, which runs through the western suburbs of the city. The appearance of this earthquake fault demonstrated for the first time that the fault continued northwards from San Francisco. Previously it had only been traced for about 600 km southwards from the city.

The 1906 earthquake was widely thought to have been caused by fault movements bearing in mind the fact that the fault moved suddenly, but in 1911 Reid proposed a theory of elastic rebound for the mechanism of earthquake generation and fault movements, on the basis of precise measurements taken around the fault. The single couple model of force he proposed was accepted as the focal mechanism, but was replaced in the 1960s with a double couple model. His elastic rebound theory, however, still survives as a fault = earthquake mechanism.

The event of 1906 in which an ordinary fault produced fault movements, gave rise to the concept and the term '*active fault*'. Geomorphologists still come to visit the distinctive topographical features seen along the fault as an aid to understanding the topography formed by an active strike-slip fault.

The attention of geologists was drawn to the fact that displacement of the fault at the time of the 1906 earthquake took the form of horizontal slippage. Subsequent geological studies revealed that the fault had brought about horizontal displacements of several kilometres or more in the rocks on both sides throughout geological time. In 1953 Hill and Dibblee (1953*) stated that the extent of this displacement had been more than 500 km since the Cretaceous. At about the same time a theory was proposed claiming that rocks on both sides of the *Alpine Fault* in New Zealand had become displaced horizontally by about 450 km. The 1950s was the decade in which geologists everywhere started to pay attention to such large *strike-slip or lateral faults*. The paper by Moody and Hill (1956*) which claims strike-slip faults to be the basis of the world's geological structures is typical of this. In the 1960s the San Andreas fault came to be regarded as an example of transform faults (Wilson, 1965*) and became a touch-stone for the theory of plate tectonics.

To call the San Andreas fault 'active' did not imply that it moved a little every day. Rather, it meant that one day it would probably move as it did in 1906. However, an area in southern San Francisco was subsequently found in which the fault is active in the literal sense and the movement is continuous. Cracks appeared in the floor and walls of a winery located immediately over the fault even when no particular earthquake activity was apparent. In 1960 these disconcerting phenomena were recognized as movement of the fault and reported to the academic world. It was the San Andreas fault that taught geologists that such continuous fault movement can really exist as one type of fault movement. The phenomenon was named '*tectonic creep*' and it was later also observed in the North Anatolia Fault Zone in Turkey.

The San Andreas fault and its activity has thus had a considerable influence on many branches of the earth sciences. In this chapter we intend to concentrate mainly on its geological properties.

(a)   Distribution and structure of the fault

Figure 2.II.1 is a map showing the broad outline of the San Andreas fault. It runs southeast almost in a straight line from Point Arena, about 160 km north of San Francisco and passes San Francisco to the southwest. It then cuts through the Coast Ranges and crosses through the Transverse Ranges to reach the low land around the Salton Sea. In the north it enters the sea at Point Arena and changes its course to virtually east-west near Shelter Cove to the south of Cape Mendocino, and continues into a large shear zone (the *Mendocino fracture zone*) on the floor of the Pacific Ocean. At its southern

Figure 2.II.1 Principal faults in California (mainly after Dickinson and Grantz, 1968* and Jennings, 1965*). Thick lines are fault lines. Thick broken lines are inferred sections. SA: San Andreas Fault. B: Big Pine fault. C: Calaveras fault. G: Garlock fault. H: Hayward fault. MF: Mendocino fracture zone. N–I: Newport–Inglewood fault. SG: San Gabriel fault. SJ: San Jacinto fault. Fine broken lines are boundaries of topographical provinces

extremity it enters Mexico and then links up with the *East Pacific Rise* which comes up from the south in the Gulf of California. Its length over land alone (from Shelter Cove to the north shores of the Gulf of California) is about 1300 km. Its course on a map is generally northwest – southeast (N40–50W) but in the northern Transverse Ranges, north of Los Angeles, its course becomes almost due east–west and the line of the fault bends markedly. Several other large faults running NE–SW (the Garlock and Big Pine faults) are also found in this area, and its geological and topographical structures both become more complex. This section of the fault is known as the Big Bend. To the north and south of the bend not only does the general strike of the fault differ, but to the south it branches into several remarkable faults. The amount of displacement of the geological terrains along the fault has clearly been less in the south than in the north.

To the immediate northwest of the Big Bend lies the famous Carrizo Plain, a semi-desert intermontane basin. Some beautiful examples of topographic features associated with faulting are found along the northeast margin of the plain following the fault line. Further northwest it emerges in the lowland around San Francisco Bay running through the valleys between the Diablo and Gabilan Ranges. The Calaveras and Hayward faults branch off to the north from here. The town of Hollister is near this point, and stone walls along its streets are distorted because of the creeping displacement. To the northwest of Hollister the fault cuts across the hills bordering the western side of the San Francisco Bay lowlands and then travels further north across the sea floor, some 10 km west of the Golden Gate. San Francisco's international airport is located only a few kilometres east of the San Andreas fault. During landing or takeoff one may see spectacular linear fault topography and the San Andreas Lake, which lies on the fault and gives it its name.

In southern California to the south of the Big Bend, the San Andreas fault branches into the Banning and Mission Creek faults, to the east of Los Angeles. In addition (to the west), other faults such as the San Gabriel and San Jacinto run almost parallel. The Salton Sea, whose eastern margins are crossed by the San Andreas fault, is a long thin strip of land below sea level and contains many fault-related features such as small volcanic cones and hot springs. This lowland continues southwards into the Gulf of California.

As mentioned earlier, the San Andreas fault is accompanied by a number of similar faults running almost parallel to it. These are usually considered together and called the 'San Andreas fault system'.

Although the San Andreas fault is shown as a single line in a diagram on the small scale of Figure 2.II.1, a more detailed map such as 1:250 000 or 1:50 000 would show it to be an assemblage of several fault lines. As a whole these form a fault zone several kilometres wide (the fault system described above is composed of a combination of fault zones). A number of lens-shaped fault slices are found inside the fault zone (Figure 2.II.2). The material of which they are composed frequently differs from that of the surrounding rocks, and they developed as fault movement caused the rock on both sides of the fault to separate and migrate. The development of this type of fault zone is thought to be due to a slip plane (fault plane) formed in a rock being abandoned for some reason, and the formation of new slip planes nearby. Generally speaking the strike of a fault active in an early period will have a trend not

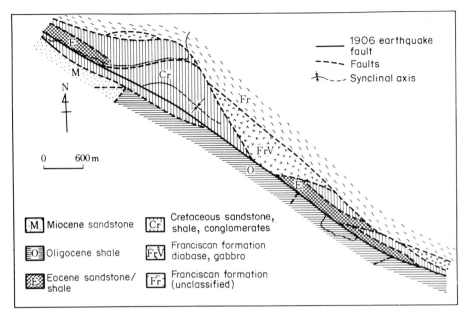

Figure 2.II.2 Example of the San Andreas fault zone (vicinity of Santa Cruz mountain. From Rogers, 1973*)

exactly parallel to the general strike and may be considerably curved. In contrast, fault lines active in the Quaternary are relatively straight. On the basis of these facts, there is an idea that the old faults have formed in *en échelon* arrangements, during a later phase of movement they become connected, and in the latest stage a smooth fault line emerges. However, there is another theory which attributes the differences to mechanical non-uniformity in rocks adjacent to a fault as shown in Figure 2.II.3 (Rogers, 1973*). This envisages a sequence in which localized plastic deformation of the rocks takes place due to their differing properties. This results in bending of the original fault line, then an increase in frictional resistance in the curved section and finally the formation of a new and straight fault line with relatively low frictional resistance. Additionally, there may be some collapse and falling in of sedimental layers which were deposited on the fault zone as a result of a vertical displacement of components accompanying the lateral displacement. At all events the San Andreas fault possesses a well developed wide fault zone suggestive of a complex history.

Rocks in the vicinity of the fault plane have often been sheared, crushed and fractured both macroscopically and microscopically by its movement. Such sheared rocks have been treated under the general title of '*cataclastic rocks*'. When the shearing movements along the fault occur at relatively large

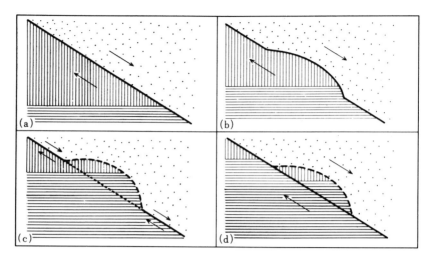

Figure 2.II.3   Diagram illustrating the development of fault zone (plane) (from Rogers, 1973*). When rocks with differing properties come into contact (a), the boundary (fault plane) between them deforms (b), (c), and finally a new straight fault plane forms where resistance is least (d). Solid line = fault line in active period. Broken line = inactivated fault line

depths under high confining pressures, the rocks apparently remain solid, but microscopic examination reveals them to be pulverized internally. Under conditions of low confining pressure crushed rocks tend to become clayey and are called 'fault gouge' or 'fault pug'. Such fault gouge is said to be well developed along the fault lines active during the Quaternary in the San Andreas fault zone.

It may be concluded from observations of fault planes within the fault zone and the linear distribution of the zone, that the dip of the San Andreas fault as a whole is almost vertical. Detailed seismic studies have shown micro earthquakes to be distributed in a planar fashion underground following the fault zone, and that this plane is almost vertical. The generation of these micro earthquakes is limited to a depth of 10–20 km or less. No earthquakes occur at greater depths than this, and it is very likely that the relative displacement of the two sides at depth is eliminated by plastic deformation.

(b)   Fault movement in Tertiary and pre-Tertiary times

In 1953 Hill and Dibblee published an important treatise on the San Andreas fault. Combining Dibblee's long experience of geological surveys, and data available up to that time, they concluded that the older the strata along the fault, the greater would be their right–lateral displacement, this amount being as much as 500 km for Cretaceous strata. Information on the age and degree of displacement of the various strata subsequently became more accurate, and nowadays virtually no one disputes the occurrence of a right-lateral displacement of 300 km or more from the Miocene to the present day.

Much work has been carried out into the displacement of strata from each period during the Tertiary and Cretaceous (Figure 2.II.4). The most abun-

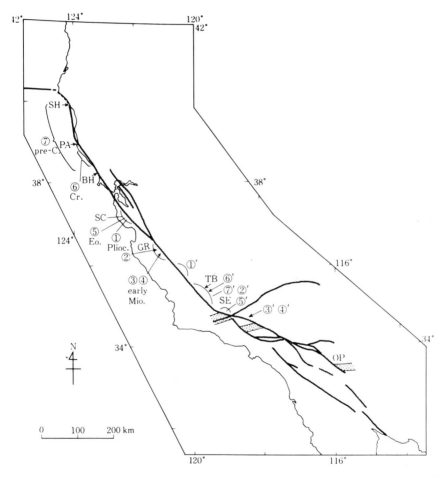

Figure 2.II.4   Present positions of rocks displaced by the San Andreas fault. (1)–(7) and (1¹)–(7¹): corresponding strata and rocks from each period. Numbers correspond to those in Figure 2.II.5. (1) Pliocene; (2) Upper Miocene; (3) and (4) Lower Miocene; (5) Eocene; (6) Cretaceous; (7) Pre-cretaceous basement rocks. Stippled areas: basement rocks including Miocene and older formations in southern California. SH: Shelter Cove. PA: Point Arena. BH: Bodega Head. SC: Santa Cruz Mts. GR: Gabilan Range. TB: Temblor Range. SE: San Emiglio Mts. OP: Orocopia Mts.

dant and reliable data relates to displacement in the Miocene rocks. Marine- and continental-formed strata from the various phases of the Miocene are found widely on both sides of the fault. The ancient geographical features of these strata, such as the shapes of sedimentary basins, thickness and distribution of the sediments, sedimentary facies, in particular the distribution of marine- and continental-formed strata which provide information about the ancient coastline, together with the distribution of fossil faunas or characteristic pebbles or sands contained in the sediments, are all unnaturally discontinuous along the line of the fault (Addicott, 1968*; Huffman, 1972*). If one brings these back along the fault line, and matches them, Miocene volcanic rocks to the east of the Big Bend correspond with an area of similar Miocene volcanic rocks in the Gabilan Range south of San Francisco. The volcanic rocks not only resemble each other in terms of their petrological characteristics and stratigraphical sequences but have also been found to be identical in terms of radioactive age and trace element components (Matthews, 1973*). This research established with virtual certainty that there has been a right–lateral displacement of about 310 km since 23.5 Ma to the present day, about 295 km since 22 Ma and 240 km since 8–12 Ma.

Attempts at reconstructing ancient geography have been made for the Eocene and Cretaceous strata also, and a right–lateral displacement of about 305 km has been deduced since 44–49 Ma (Clarke and Nilsen, 1973*) and of about 500 km since the deposition of Cretaceous strata. It was noticed that the value of approximately 305 km since 44–49 Ma was almost equal to the value of approximately 310 km since 23.5 Ma within the range of error. Values for pre-Cretaceous periods have been inferred from apparent displacements of the distribution between pre-Cretaceous granitic basement rock (Salinian blocks) on the west side of the fault and similar basement rocks on the east (approximately 500 km +) but the precise values are obscure. This is because the northern limits of the Salinian blocks on the west beyond Bodega Head, 70 km north of San Francisco have still not been identified with certainty, as is also the case with the position on the east side from which they migrated. However, the results of recent studies of Sr isotope ratios in Salinian blocks have suggested a displacement of about 510 km very much in accordance with the estimation so far.

Figure 2.II.5 shows the displacements of rocks from various periods. This graph implies that there was little activity along the San Andreas fault between 50 and 20 Ma (Eocene early Miocene). It became active again at some point between 20 and 10 Ma and this has subsequently continued to the present day with an accelerating rate of displacement.

Virtually all the data discussed above were obtained from the area north of the Big Bend. To the south of the bend, studies are much more complicated owing to the development of parallel or even left-lateral faults almost at right angles to the main fault each possessing its own history (Crowell, 1973*).

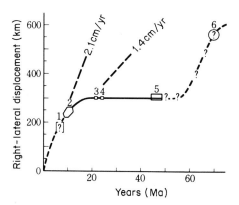

Figure 2.II.5 Changes in displacement rates along San Andreas fault (from Clarke and Nilsen, 1973*). Numbers 1–6 in the diagram show age of separated and slipped rocks (horizontal axis) and amount of displacement (vertical axis). Slope of line joining each point expresses average rate of displacement in that period. 1–6 corresponds to (1)–(6) in Figure 2.II.4

However, it should be noted that only right-lateral displacement of about 300 km since the deposition of Miocene formations has been confirmed south of the Big Bend and no proof of displacement prior to this has been obtained. In southern California the Miocene formations found to the southwest of the Big Bend (near Tejon) have migrated about 260 km south in total (to the Orocopia Mountains) together with pre-Tertiary basement rocks along the San Andreas fault and the San Gabriel fault running parallel to the west (Crowell, 1962*, 1973*). As pre-Tertiary basement rocks containing some Precambrian rocks can be compared in both areas, the beginning of activity in these faults must have been during or after the deposition of Miocene formations (about 12 Ma).

To sum up, the origins of the San Andreas fault in southern California seem to be fairly recent and its total displacement is also barely half that observed north of the Big Bend (500–600 km). Many people believe therefore that faults other than the present San Andreas fault were once active in Southern California, and that these account for the missing 200–300 km of displacement. For example, Suppe (1970*) thought that the Newport–Inglewood fault near Los Angeles (see Figure 2.II.1) was a continuation of the San Andreas fault north of the Big Bend in the Palaeogene time, and the missing 300 km displacement occurred there. Suppe named it the 'proto-San Andreas fault' and devised a reconstruction in which he moved the pre-Cretaceous Salinian blocks on the west southwards relative to the eastern side along this fault (see Section 2.VI, Figure 2.VI.2).

(c)   Quaternary fault movements

We mentioned earlier that part of the San Andreas fault is currently subject
to continuous movement. Detailed measurements indicate an annual average
rate of a few centimetres (5 cm or less) although it varies dependent on place
and time. Over the past 60 years the average rate in southern Hollister, as
inferred from the offset of old fences on farms and so on, has been no more
than 2 cm per year. This type of creeping fault movement is not found at all
further south in the Carrizo plain area or around the Big Bend. However
much topographical evidence such as the bent configuration of valleys, stream
offset and displacement at the time of the great earthquake of 1857
(right-lateral displacement approximately 10 m) suggests that fault displacement occurs in these areas only when there is a major earthquake, such as in
1857, which occur every few hundred years. If such infrequent severe
displacement in association with earthquake is averaged over time, the rate of
displacement along the fault still comes out at 2–4 cm per year, very similar to
the areas of creeping displacement.

Such rates of displacement are less than the rate of horizontal slippage
(approximately 5 cm per year) expected from horizontal strain rates in the
fault area detected by geodetic methods, and also less than the value for
relative velocity between the *Pacific* and the *America plate* deduced from the
rate of sea floor spreading in the Gulf of California (approximately 6 cm per
year). As we discuss later, this is thought to be because the San Andreas fault
is affected by only part of the relative displacement between the two plates,
and the rest is absorbed as displacement along other faults and crustal
deformation over a wide area on the western margin of the American
continent from Western California across the Sierra Nevada to the Basin and
Range Province in the east.

If a geological survey shows strata of different age meeting at a fault, we are
apt to assume this to be the result of upwards or downwards displacement of
the basements on both sides of the fault. However, such a distribution can be
produced without any upwards or downwards displacement at all. The reason
is that because strata are not unlimited in the horizontal direction and
moreover are not horizontal, they may well come up against strata of a
different age simply as a result of lateral displacement. The 'lateralists' have
frequently pointed this out in relation to the history of the San Andreas fault
(Hill and Dibblee, 1953*; Crowell 1962*).

If one looks at the topography along the San Andreas fault, there are
significant indications that vertical displacement has occurred in certain areas
at least in the Quaternary. However, it is possible to say that this fault is an
almost perfect macroscopic example of a strike-slip fault of very long
standing, because, looking across the immense tracts of geological time,
strata formed in almost identical sedimentary environments in the same

period are still found to be of about the same height even though separated horizontally by 300 km or more.

Many large and small depressions and elevations have been produced along the line of the fault as a result of movements during the Quaternary time. If one traces such features along the fault line, one can readily notice that the sense of the vertical displacement changes within a short distance. In the Carrizo Plain, for example, long narrow hills following the fault line, produced by the relative elevation of the southwest side of the fault line gradually become low for several hundred metres with a considerable gradient in the direction of strike, but the northeast side conversely becomes upheaved. Graben-like depressions are often found on the fault line at the foot of such hills, but over a short distance these also become shallow and thin and vanish into the hillsides. The origins of such alternating elevations along an almost perfect strike-slip fault are thought to lie in the fact that when there is lateral movement along a fault plane which is not perfectly smooth in geometric terms, localized extensions and compressions will be set up in curved areas of the crust which will produce depressed and elevated surface features, respectively. Much work has been undertaken in New Zealand into the fact that the alignment of such vertical displacement along the line of the strike-slip fault is uniform for neither location nor time, and this is regarded as one particular feature of strike-slip faults.

(d)   The San Andreas fault as a plate boundary

Maps of plates across the world represent the San Andreas fault as the boundary between the Pacific and the American plates. The banded pattern of magnetic anomalies on the Pacific floor off California to the south of the *Mendocino fracture zone* suggests that the age of the ocean floor decreases the nearer it approaches California. Therefore the ocean ridge which produced this sea floor must now have disappeared underneath the American continent. The Gorda and Juan de Fuca ridges off northern California and the *East Pacific Rise* which extends as far as the Gulf of California from the south may be thought to be remnants of that ridge. In this sense the San Andreas fault is a transform fault connecting two north and south ocean ridges (Wilson, 1965*; Atwater, 1970*).

The age of the part of the ocean floor in contact with the American continent off California is greatest (29 Ma) off Cape Mendocino in the northern part of the San Andreas fault. It becomes progressively younger to the south being only about 4 Ma in the Gulf of California in Mexico. Therefore, it is thought that the ocean ridge, which made this floor, came into contact, from the west, with the trench *subduction zone* off California near Cape Mendocino at around 29 Ma, and was absorbed by that trench and disappeared under the American continent. At that time the directions of the ridge (almost N–S) and trench (NW–SE) were not parallel (Figure 2.II.6), and so the ridge disappeared from the north. As a result, the trench became a transform fault (the San Andreas fault). (In the geometry of *plate tectonics*, this must happen in the situation shown in Figure 2.II.6 (see vol. 1, Chapter 5). In this way the transform fault spread southwards replacing the oceanic trench and reached the Gulf of California about 4 Ma ago.

Such inferences from the oceanic plate imply that the San Andreas fault was born and displacement began at around 29 Ma. The southwest side of the fault should also be an oceanic plate. However, neither point harmonizes with the geological data for the continent discussed previously. How can they be explained? The explanation proposed by Atwater (1970*) and Garfunkel (1973*) runs as follows. The transform fault which began to develop off California since 29 Ma was not itself the San Andreas fault. The predecessor of the present fault existed on the American continent before 29 Ma and was

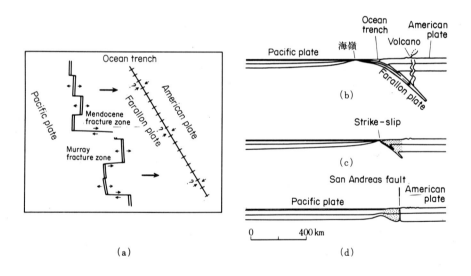

(a)   (d)

Figure 2.II.6   Changes in the boundary between the Pacific and American plates off California (from Atwater, 1970*). (a) Situation before Pacific plate collided with American plate (approx. 53 Ma). An ocean trench was present and Farallon plate was subducting along it. (b) Profile of same. (c) Situation when ridge producing Pacific plate reached trench off California (approx. 29 Ma). Trench changed to transform fault. (d) Eastern margin of Pacific plate consolidated with American plate, and lateral displacement began to occur mainly in the continental interior (present San Andreas fault, etc.) (from later Miocene to present). Stippled areas show continental block west of San Andreas fault

slipping right-laterally. After 29 Ma the land block (stippled areas in Figure 2.II.6(c) and (d) between the above newly-formed transform fault (strike-slip fault in Figure 2.II.6(c)) and the existing San Andreas fault, gradually bonded with the offshore transform fault and began to move with the Pacific plate. The relative displacement of the American plate then mainly occurred along the eastern margins of this block, namely the present San Andreas fault. From the Miocene onwards the rate of right-lateral displacement of the San Andreas fault accelerated (Figure 2.II.5) due to the increased degree of bonding over time between the offshore transform fault and the continental block on its eastern margin. As the time when the offshore oceanic trench changed to a transform fault was immediately after it had absorbed the ridge, the plate boundary was still relatively hot and soft and slipped laterally along the trench axis. With the passage of time, however, it cooled and solidified, making movement more difficult, so that displacement came to take place chiefly along the existing weakness on the continent, namely the San Andreas fault.

In essence, therefore, the broad picture of movement in the San Andreas fault at least after the mid-Tertiary time is compatible with the characteristics of relative displacement between two plates, the American and the Pacific, which form part of the worldwide plate system.

Several other major *strike-slip faults* in the 1000 km class of the San Andreas fault are known on other continents. The majority of these are active and are well recorded topographically in satellite pictures. The principal examples in the Circum-Pacific belt are the Denali fault system in Alaska (approx. 2000 km long, right-lateral displacement 400–700 km), the Median Tectonic Line in Japan (approximately 1000 km long, right-lateral), the Philippine fault zone (approximately 1300 km long, left-lateral), the Great Sumatra fault zone in Sumatra (approximately 1800 km long, right-lateral), the *Alpine fault* in New Zealand (approximately 1000 km long, right-lateral displacement approximately 450 km), the Atacama fault in Chile (approximately 800 km long, right-lateral), etc. In Eurasia one may cite the Altyn Tagh fault (approximately 1500 km long, left-lateral) in China, together with the Talas-Fergna fault in the Kirghiz-Kazakh region of the USSR (900 km long, right-lateral displacement 250 km), the Herat fault (1100 km or more long, right-lateral), and Chaman fault (800 km long, left-lateral displacement 500 km), and the North Anatolia fault in Turkey (900 km long, right-lateral).

The majestic sharp straight lines carved out in the surface of the earth and revealed by satellite pictures are these faults. One of the tasks of the Earth Sciences must be to try and explain the origins of these wrench faults with their lateral displacements of hundreds of kilometres.

Tokihiko Matsuda

References

Addicott, W. O. (1968): Mid-Tertiary zoogeographic and paleo-geographic discontinuities across the San Andreas fault, California, *Stanford Univ. Publ., Geol. Sci.*, **11**, 141–65.

Atwater, T. (1970):Implications of plate tectonics for the Cenozoic tectonic evolution of western North America, *Geol. Soc. Am. Bull*, **81**, 3513–35.

Clarke, S.H. and Nilsen, T.H. (1973): Displacement of Eocene strata and implications for the history of offset along the San Andreas fault, Central and Northern California, *Stanford Univ. Publ. Geol. Sci.*, **13**, 358–67.

Crowell, J. C. (1962): Displacement along the San Andreas fault, California, *Geol. Soc. Amer. Spec. Paper*, **71**, 1–61.

Crowell, J. C. (1973): Problems concerning the San Andreas fault system in southern California, *Stanford Univ. Publ., Geol. Sci.*, **13**, 125–35.

Dickinson, W. R. and Grantz, A. (eds) (1968): Proc. Conf. Geol. Problems San Andreas fault system, *Stanford Univ. Publ., Geol. Sci.*, **11**, 374 pp.

Garfunkel, Z. (1973): History of the San Andreas fault as a plate boundary, *Geol. Soc. Am. Bull.*, **84**, 2035–42.

Hill, M. L. and Dibblee, T. W. (1953): San Andreas, Garlock, and Big Pine faults, California, *Geol. Soc. Am. Bull.*, **64**, 443–58.

Huffman, O. F. (1972): Lateral displacement of upper Miocene rocks and the Neogene history of offset along the San Andreas fault in Central California, Geol. Soc. Am. Bull., **83**, 2913–46.

Jennings, C.W. (1975): Fault map of California, 1:750,000, California Div. Mines Geol., *Geologic Data Map, No. 1*.

Matthews, V. III (1973): Pinnacles-Neenach correlation: a restriction for models of the origin of the Transverse Ranges and the big bend in the San Andreas fault, *Geol. Soc. Am. Bull.*, **84**, 683–8.

Moody, J. D. and Hill, M. J., (1956): Wrench-fault tectonics, *Geol. Soc. Am. Bull.*, **67**, 1207–46.

Rogers, T. H. (1973): Fault trace geometry with the San Andreas and Calaveras fault zones—a clue to the evolution of some transcurrent fault zones, *Stanford Univ. Publ., Geol. Sci.*, **13**, 251–8.

Suppe, J. (1970): Offset of late Mesozoic basement terrains by the San Andreas fault system, *Geol. Soc. Am. Bull.*, **81**, 3253–8.

Wilson, J. T. (1965): Transform faults, oceanic ridges and magnetic anomalies southwest of Vancouver Island, *Science*, **150**, 482–5.

III *Fault block movements in Shandong Province, China*

Almost the first thing one notices when looking at a world map is the Alpine–Himalayan mountain system which cuts through the continent of Eurasia from east to west. Although most areas on the northern side of these ranges represent stable geomorphological features, extraordinarily high relief prevails eastwards from the Pamir plateau over almost the whole of the Chinese continent. Seismic charts indicate that many large earthquakes of shallow origin have occurred, closely related to this tectonic geomorphology. Many studies of the Pamir and Tien Shan regions, carried out in the Soviet Union, have suggested that after once becoming a part of the continental crust and stabilizing, these areas have become active again, since the Cenozoic. Particularly remarkable *fault block movements* have become prominent in the Quaternary period to produce the mega-tectonic topography seen today. Terms such as 'neotectonics' or 'restored mountain range' were derived from such studies.

After the revolution of the New China in 1949, it was believed that rapid strides were being made in geological survey work, but no details of the work were forthcoming. In 1975, however, a 1:4 000 000 *Map of Geologic Structural Systems of China* was published, followed in 1976 by a 'Geological Map of China' on the same scale. These clarified the outline of the structural framework of the country and led to the conclusion that the fundamental structures of China are of the '*Kuaidan*', or fault block type (Tectonic Map Compiling Group, 1974*). Thus the origins of the mega-relief seen in the Chinese continent are understood to lie in differential movements of blocks bounded by faults (Inst. Geomechanics, 1976*).

Such fault block movements are very closely related to seismic activity. In 1966, research, which was looking at earthquake prediction following the Xingtai (Singtai) earthquake in Hopei Province, revealed a good correlation between fault structures and the distribution of many historical, large earthquakes, recorded on priceless documents spanning more than 3000 years (Figures 2.III.1). Then in 1972, images sent back from the LANDSAT/ERTS satellite threw the huge Chinese active fault systems into relief and led rapidly to a new appreciation of fault block movement (Molner and Tapponnier, 1975*; York *et al.*, 1976*).

It scarcely needs saying, however, that the fault blocks seen today are the result of all the different vertical crust movements and the changes of stress field that have taken place since Precambrian times. In this section we discuss the history of the tectonic development of the fault block structures of Shandong Province, as an example of the fault block movements of China. The reasons for this are that relatively detailed field surveys have been published for this region and correlations have also been made with the recent Xingtai, Haicheng, Bohai and Tangshan earthquakes. The area is also thought to be closely connected both structurally and locationally with the western half of the Japanese Islands.

(a)   The position of the Shandong geological province from the viewpoint of the tectonics of China

The Shandong geological province lies between the great northern alluvial plain, where the Huang Ho (Yellow River) flows into Bo Hai (Gulf of Chihli) and the southern plains, where the Huai Ho and Chaug Jang (Yangtse Kian)

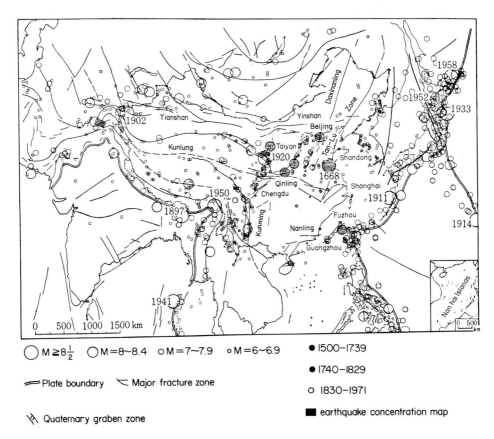

Figure 2.III.1   Tectonic zones and distribution of major earthquake epicentres in and around China (Shi, *et al.*, 1973*)

enter the Yellow Sea. It extends almost to the line of Jinan and Xuzhou and centres on the Shandong Peninsula and Taishan (1524 m). It is about 700 km across in an east–west direction and about 400 km in the north–south direction.

Although recently a new concept of intraplate tectonics has been introduced among the Chinese earth scientists, the foundations of the consideration of the geological structures of China were mainly laid in the geodynamic studies of Li Siguang. These were succeeded by the Institute of Geomechanics in Beijing (Peking) and became the cornerstone of structural geology in China. *The Map of Geologic Structural Systems of China* may be said to be a summary of these works. It attempted to make a broad classification of all the complex structural elements in terms of both their alignment and period. The most important categories identified were (1) the so-called *East–West (latitudinal) tectonic zones* and (2) *Neo-Cathaysian tectonic zones*. Group (1) are mainly compressed structural zones, formed in the period of Variscan orogeny in the late Palaeozoic. They comprise several zones which can be traced for up to 2000 km and form the geological framework of China. Their most northerly element is the Tien Shan – Yin Shan zone. In the centre lies the Kunlung-Qinling zone and their southern component is the Nanling zone. A particular feature of eastern China is the younger structures of group (2) superimposed on these pre-existing ancient structural belts. The newer structures are represented by the Dahinanling (Great Khingan Range), etc. running in a NNE–SSW direction and taking the form of large scale warp structures. They became prominent after the Yanshan movements in the late Mesozoic, producing depressed areas in the alluvial plains, the Yellow River and Bo Hai, and lifted areas in the Liangdong and Shandong peninsulas.

By contrast, it is noticeable that more recent structures in western China such as in the Xizan (Tibet) or Xinjiang (Sinkiang) areas are also aligned in an east–west direction. Therefore Sun Dianging (1966*) postulated that 'There have been two principal directions of movement of the Chinese continent since the Yanshan Movements. The first was compression in a north–south direction, and the second was compression in an east–west direction. Because the two series of compressive movements were unequal, this resulted in torsional pressures forming twisted structures. The form of movement common to all these types of structure was horizontal.' Even today, the interpretation has not been changed. The close relationship between these structures and earthquake activity is vividly demonstrated in Figure 2.III.2 and major earthquakes may broadly be said to occur in the zone of torsional zones connecting Peking, Taiyuan, Xian (Sian), Cheug du and Kunming. The Shandong geological province described here lies between the Tien Shan – Yin Shan and Kunlun-Qinling zones and has been strongly affected by structural movements of the Neo-Cathaysian system. The Tectonic Map Compiling Group (1974*), from the Geological Institute of the Chinese Academy, summarized the history of the development of Chinese geological structures, and regards the Tan-Lu fracture zone, which crosses the central part of the Shandong geological zone as particularly significant to the seismology of the area. The maps explain the relationship between earthquakes of the northeastern China and geological structures, by combining this with the east–west structural zones.

(b)   Neotectonics of Tan-Lu fracture zone

The most remarkable structure in eastern China is known as the Tancheng–Lujiang fracture belt running NNE–SSW (this is commonly abbreviated to Tan-Lu, as is the case below). It is most obvious at the root of the Shandong

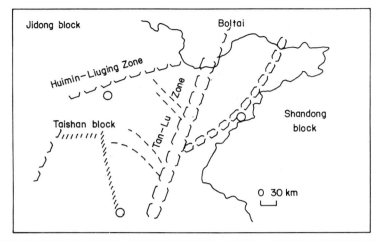

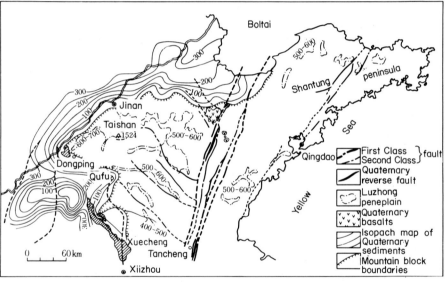

Figure 2.III.2   Recent structures and topographical planes of Shandong geological
province (Study Group for Shandong Seismogeology 1974)

Peninsula, in an area also known as the Yi Shw deep fracture zone. Following
it to the north, it enters Bo Hai Bay, but appears again west of the Liaodong
Peninsula. It then extends northwards along the Songhuajang (Sungari river)
as far as the vicinity of Jiamsi. To the south it sinks beneath the alluvial plain
of the Huei Ho, as the Shandong hills become lower, and is assumed to
extend as far as the vicinity of the river Chang Jiang (Yangtse). This fault

system, which in total covers more than 2000 km, is the most important structural zone in eastern China belonging to the Neo-Cathaysian system. Recent activity centres on its central section particularly the Gulf of Bo Hai between the Liaodong and Shandong Peninsulas.

This province is divisible into two huge blocks by the Tan-Lu zone. We call these the Shandong block to the east, and the Taishan block to the west. In Chinese literature the former is known as the Jiao Liao block and the latter as the Luzhong block, but we decided to use more familiar names for the sake of non-Chinese readers. Taking the Tan-Lu zone as the first order fault, the two blocks are further divided by the second order faults as shown in Figure 2.III.2. This is a pattern typically encountered in Chinese fault block structures.

Much work has been done in the past on the ancient rocks found in this province, but recently the tectonic studies have made rapid progress concerning the earthquakes, which frequently occur around the Gulf of Chihli. So in this section I would like to look principally at the reports of the Geological Institute of the Chinese Academy.

As mentioned earlier, the Shandong geological province can be divided into the Shandong and Taishan blocks by the Tan-Lu zone. The western Taishan block is divided from the block (Jiton – Bo Hai block), occupying the alluvial plain of the Huang Ho (Yellow River), by a structural zone running almost due east–west. This is known as the Huimin-Linging fracture zone. Although there is a predominance of faults striking NW or WNW in the Taishan block, NE or ENE faults have developed in the Shandong block to the east. When combined with the NNE fault system of the Tan-Lu zone, they form a framework of structures as the basement fault system of the region.

We initially describe the newer fault movements of the area south of Tancheng, according to survey by Fan Zhongjing *et al.* (1976[*]). This region extends from the Shandong mountain area to where the Tan-Lu zone gradually sinks beneath the alluvial plains, and also demonstrates various topographical features. Cretaceous shale, sand and conglomerates together with Cenozoic sand and conglomerate layers are arranged in bands whilst forming isolated monadnocks with relative heights of 35 – 120 m.

This situation is illustrated in the block diagram in Figure 2.III.3. Geological field surveys, drilling and geophysical researches in the northern area have revealed that these monadnocks are bounded by four concealed basement faults ($F_1 - F_4$). Early and late Cretaceous formations are distributed between $F_1 - F_2$ and $F_3 - F_4$ burying graben sections of the Precambrian gneiss. It appears that folding of these Cretaceous formations also occurred, accompanied by thrusting, and the anticlinal parts of the folds and thrust blocks protruded above the alluvial plain to form the island-like monadnocks.

Despite discontinuities, such thrust faults may be followed for 200 km and can be confirmed at many outcrops. Their displacements are shown in Figure

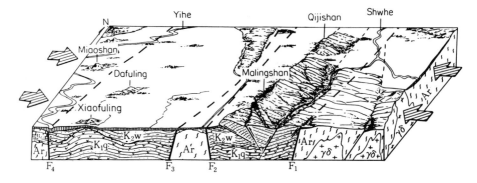

Figure 2.III.3   Geological section of the central part of Tan-Lu zone. Stress fields and block diagram showing structures around Malingshan south of Tancheng (Fan, *et al.*, 1976*). F: basement faults; Ar: Archaean gneiss; γ^{∂}: Mesozoic granites; K_1q: early Cretaceous formations; K_2w: late Cretaceous formations; Arrows: direction of compression

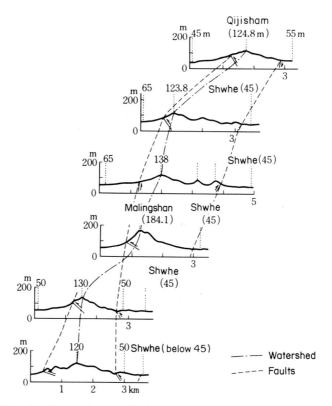

Figure 2.III.4   Profiles and fault displacements in the central part of Tan-Lu zone (Fan, *et al.*, 1976*).

2.III.4. Figure 2.III.5 is a sketch of a typical outcrop on the Xiaoshang-zhuang-Hezhuang fault running to the west of Mount Malingshan. The later Cretaceous sandstone to the east, covered by the early Quaternary conglomerate is thrust over the mid and late Quaternary conglomerate beds along the line of the fault. The dip of the fault plane is about 50° in this area, but varies between 50 – 75°. It should be noted that although new fault gouge and breccia are about 0.5 m wide, the adjacent crushed zones of later Cretaceous sandstone reach 20m or more. This implies that movements did not take place only in the late Quaternary, but that there had also been repeated fault movements in earlier periods.

Widespread peneplanation in the Shandong geological province prior to the Quaternary may also be inferred from the lack of Neogene formations. The elevated peneplain, found on the summits of the mountains, called the Luzhong peneplain, was formed in the early Tertiary. It was uplifted from the end of the Tertiary and affected by severe fault block movements, presently being found at altitudes of 400 – 700 m (Figure 2.III.2). The plain one step lower than this at 300 – 400 m is covered by the late Pliocene Tangshan conglomerates and basalts, and forms hills 180 – 200 m high, near the coast of the northern Shandong peninsula. This is the Tangshan plain. The third Lincheng plain forms river terraces 150 – 200 m high, in the mountains and coastal terraces 100 – 200 m high. The details of these plains cannot be clearly understood simply from a description, but when compared with the topographical plains of Japan, the Luzhong plain may be seen as corresponding to the Chugoku peneplain in Southwest Japan, the Tangshan plain to the Kibi peneplain and the Lincheng plain to the terrace surface.

Deformation and displacement of these plains indicate that, although the entire Shandong province rose gently throughout the Tertiary, the primary peneplain was maintained as a result of the balance between denudation and elevation. However, in the Quaternary the Shandong block began to suffer tilting movements accompanying warping, to produce the present topogra-

Figure 2.III.5   Sketch of outcrop on Xiaoshangzhuang-Hezhuang fault running to the west of Malingshan to south of Tancheng. (Fan, *et al.*, 1976*). Q: Quaternary formation; K_2w: late Cretaceous formation

phy. In contrast, whilst the Taishan block rose it was split into further blocks by NE striking faults. There was also a relative subsidence on the south-west side of a line connecting Dongping – Qufu – Zuecheng, forming a Quaternary sedimentary basin (Figure 2.III.2). It is interesting to note that this basin corresponds with that of Miocene age. Then, in the middle and late Quaternary remarkable compressed structures were formed in the Tan-Lu zone as may be seen in Figure 2.III.3.

(c)   History of tectonic development prior to the Tertiary

As described above the Quaternary movements in the Tan-Lu zone were influenced considerably by the ancient pre-existing structures. Let us consider the history of pre-Tertiary tectonic development by looking at the map of basement geology in Figure 2.III.6.

In the Archaean or early Proterozoic, when the basement rocks were formed, the following discrepancies were already apparent along the fault F_1 on the western boundary of the Tan-Lu zone. The basement of the Taishan block is formed from gneissose rocks intercalated by layers deposited in the Archaean eugeosyncline, known as the Taishan Group. These structures run in a NW direction. On the other hand, the basement to the Shandong block, consists of the metamorphic rocks originating from the Jiaodong Group and corresponding to the upper Taishan Group, with the Fenzi Group resting unconformably over them. The principal structural direction of this block is NE. However, it is uncertain whether or not such differences in basement structure are connected directly to the formation of the Tan-Lu zone or not.

The area entered a relatively stable phase from about the middle of the Proterozoic. In the middle and later stages of this era, known as the Sinan Period, the Shandong region combined with the Shanxi (Shansi) region to form an elevated area, in stark contrast to the sedimentary basin of the Sinan Groups in northern China. The area around the Yi Shw deep fracture zone, at one stage, developed into a rift valley or a north–south channel linked to the sea, into which up to 100 – 200 m of Tumen Formation were then deposited. In the Palaeozoic the Shandong block continued to be in a stable state of uplift and erosion, whereas the Taishan block was affected in the Mesozoic by repeated regional transgressions and regressions of the sea (Li Guopeng and Li Yusong, 1973*).

At the beginning of the Mesozoic, the Shandong province was in a state of regional erosion, but by the Jurassic small fault basins had been created to form the basin of sedimentation. The formation of the Tan-Lu zone seems to date from about this time. This geological province then embarked on a period of violent igneous activity from the late Jurassic to the Cretaceous, particularly in the Shandong block. This activity was an expression of the Yanshan movements, evidence of which remains in the acidic rocks, wide-

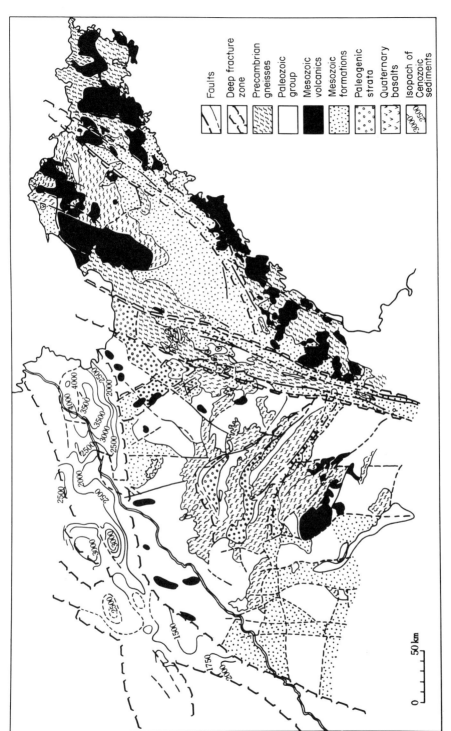

Figure 2.III.6 Basement geological map of Shandong geological region (Research Group for the Shandong Seismogeology, 1974*)

Faults

Deep fracture zone

Precambrian gneisses

Paleozoic group

Mesozoic volcanics

Mesozoic formations

Paleogenic strata

Quaternary basalts

Isopach of Cenozoic sediments

50 km

spread across a huge area of eastern China, South-east Japan and Sikhote Alin. Differences in the degree of this activity between the Taishan and Shandong blocks are quite obvious. With the occurrence of differential movement between each block, the Tan-Lu zone became clearly defined, and Cretaceous strata were deposited in its graben-like structural basin.

Pre-Mesozoic formations were consolidated as a result of the violent igneous activity during the Yanshan movements and certainly took on the role of basement rocks for the post-Cretaceous formations. Later still the basement rocks probably moved as a block, and the stresses produced were mainly concentrated in the link zone. After the Cretaceous, this area entered a phase of compressive structural movement and the Cretaceous formations deposited in the Tan-Lu rift zone began to be folded. Various other compressed structures such as schist zones, mylonite zones or crushed belts were also formed. Following on from this, the region as a whole shifted upwards, and entered a long period of denudation. The Quaternary movements are then thought to have acted on these older structures.

(d)   Characteristics of the Shandong block movements

When one looks at the history of tectonic development in the Shandong geological province described above, the following major features may be identified.

(i)   Although the currently visible Tan-Lu zone is a boundary between different geological structures, which appeared during and since the Precambrian, it was not until the Jurassic that we have clear evidence to show the incipient form of the Tan-Lu zone.

(ii)   The Jurassic to the Cretaceous was marked by the Yanshan movements. Basement rocks were consolidated as a result of violent ingeous activity, and block movement accompanied by fracturing became prominent. The Tan-Lu zone also appeared as a rift zone, into which Cretaceous strata were deposited.

(iii)   After the Cretaceous, the tectonic stress field changed from extension to compression. The Tan-Lu zone had up till then formed a wide graben zone, and its accompanying faults perhaps reached deep into the crust on a scale worthy of what is called in China a 'deep fracture zone'. When subjected to regional compressive stress, wedged between the principal blocks, the stress was concentrated, resulting in the folding of the Cretaceous formations between the blocks. Fissures produced in the Cretacious formations at virtually the same time, consisted both of those with strikes almost parallel to the principal faults, $F_1 - F_4$ and branch faults intersecting the principal faults at acute angles. It is difficult to explain all these in terms of a simple dynamic process, but we may perhaps visualize it as horizontal compressive movements accompanied by some torsional effect.

(iv)   In Cenozoic time, structural movements became much weaker, but the structural stress field, dated as the late Mesozoic, continued and controlled the formation of Cenozoic sedimentary basins. In contrast to the Huan-Huai alluvial plains, the whole of this region was uplifted, but became a peneplain as a result of long term erosion. However, from about the beginning of the Quaternary, there was a second series of strong horizontal compressive movements. The Cretaceous strata of the Tan-Lu zone were sandwiched between the basement blocks and received the full impact of the stress. Many reverse faults appeared in the latter half of the Quaternary and the Cretaceous rocks were thrust over the Quaternary sediments, and ancient crushed zones removed as they were exploited by the newer fault planes. This means that the stress environment of the late Cretaceous persisted until the Quaternary.

(v)   The occurrence of large earthquakes in this region is closely related to this fault system. A look at the earthquake mechanisms reveals the maximum principal stress axis of compression to be horizontal, crossing the Tan-Lu zone in an almost NE–SW direction. This corresponds well with the tectonic stress field.

To summarize the above, the framework for the block structures in the Shandong province was created by the Yanshan movements and developed in response to variations in the intensity of horizontal compressive movements almost at right angles to the Tan-Lu zone from the Mesozoic to the present day.

Kazuo Huzita

References

Fan, Z., Ji, F., Xiang, H. and Ding, M. (1976): The characteristics of Quaternary movements along the middle segment of the old Tancheng-Lujiang fracture zone and their seismological conditions, *Sci. Geol. Sinica.*, No. 4, 354–66.

Geological Academy of China (1975): Map of tectonic system of China (1/4,000,000), Peking.

Geological Academy of China (1976): Geological map of China (1/4,000,000), Peking.

Institute of Geomechanics, Chinese Academy of Geological Sciences (1976): On tectonic systems, Peking, 1–15.

Li, G. and Li, Y. (1973): Characteristics of the development of the Meso-Cenozoic geology of the eastern part of China, *Sci. Geol. Sinica.*, No. 3, 238–44.

Molner, P. and Tapponnier, P. (1975): Cenozoic tectonics of Asia, *Science*, **189**, 419–26.

Research Group for the Shandong Seismogeology (1974): Preliminary report on the
  Shandong Fault Blocks and seismic zone. *Sci. Geol. Sinica.*, No. 4, 315–29.
Shi, Z., Huan, W. and Cao, X. (1973): On the intensive seismic activity in China and
  its relation to plate tectonics, *Sci. Geol. Sinica.*, No. 4, 281–93.
Sun, D. G. (1966): Main systems of the geologic structure in China, *Chikyu Kagaku*,
  Nos 85–86, 69–84 (in Japanese).
Tectonic Map Compiling Group (1974): A preliminary note on the basic tectonic
  features and their developments in China, *Sci. Geol. Sinica.*, No. 1, 1–17.
York, J. E., Cardwell, R. and Ni, J. (1976): Seismicity and Quaternary faulting in
  China, *Bull. Seismol. Soc. Amer.*, **66**, 1983–2001.

IV   *Thrust faulting in the Canadian rockies*

The Canadian Rockies are a range of mountains straddling the states of
British Columbia and Alberta in a NW–SE direction. In 1885, R. G.
McConnell of Geological Survey of Canada conducted a survey of Western
Alberta, in the course of which he discovered Cambrian limestones lying on
Cretaceous shales. He concluded that a low-angle thrust fault existed between
the two (McConnell, 1887*). This fault was subsequently identified by many
researchers, and his name was given to it as the 'McConnell Thrust',
delineating the eastern margin of the Canadian Rockies (Clarke, 1949*). On
the map, the McConnell Thrust can be traced in a NW–SE direction; and the
fault plane is inclined slightly towards the southwest. Strata on the southwest
side of this fault were shifted towards the northeast and were thrust over the
younger strata.

Many *thrust faults* structurally similar to the *McConnell Thrust* are found in
the Canadian Rockies, trending parallel to each other in a NW–SE direction,
the fault planes dipping to the southwest. When and how were such low-angle
thrust faults, a particular feature in the eastern part of the Canadian Rockies,
formed? Studies of the geological structures of this area beginning with the
work of R. G. McConnell were subsequently expanded to the whole of the
Canadian Rockies, and much new geological data were obtained along with
developments in the oil and gas fields. The history of geological structures in
this mountain system is outlined as follows. From the Palaeozoic to the
Mesozoic this region was an area of widespread and gentle subsidence and a
thick sedimentary pile was formed. At the end of the Mesozoic, the
sedimentary formations glided toward the continental interior in the northeast, owing to extensive uplift in the southwestern part of the Canadian
Rockies, and were successively overthrust. The total displacement at this time
is estimated to have been in excess of 200 km. Such large scale lateral
movement accompanying elevation of the crust, the low-angle thrust faults
and folded structures formed as a result thereof, are known throughout the

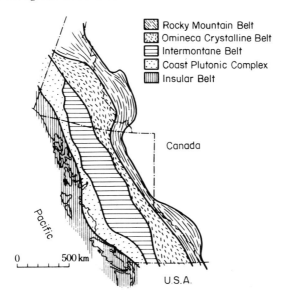

Rocky Mountain Belt
Omineca Crystalline Belt
Intermontane Belt
Coast Plutonic Complex
Insular Belt

Canada

Pacific

0   500 km

U.S.A.

Figure 2.IV.1   Physiographic divisions of the Canadian Cordillera (after Monger *et al.*, 1972*)

eastern part of the Rockies, the Western part of the Appalachian Mountains and in the Ouachita system. Similar structures are also found in the eastern part of the Andes in South America. The low-angle thrust faulting in the Canadian Rockies furnishes a typical example of geological structures commonly encountered in mobile belts.

(a) *Location of the Canadian Rockies*

A mountainous belt runs down the western part of the North American continent. Its width various from 600 – 1000 km from the Pacific Ocean to the interior plains, and it extends for some 6500 km from Alaska, through Western Canada and California to Mexico. This belt is known as the North American Cordillera.

The Canadian Cordillera is that part of the North American Cordillera in Canada. According to the physiographic features, it is divided into the following five zones from the Pacific Ocean eastwards (Figure 2.IV.1):

(1) Insular Belt
(2) Coast Plutonic Complex
(3) Intermontane Belt
(4) Omineca Crystalline Belt
(5) Rocky Mountain Belt

The Canadian Rockies are situated in the southern part of the Rocky Mountain Belt.

The Insular Belt, which includes Vancouver and Queen Charlotte Islands, is composed of Palaeozoic and Mesozoic groups, crystalline schists, and granites of various ages. The Coast Plutonic Complex is an area rich in granitic plutonic rocks, which were intruded from the end of the Mesozoic to the Cenozoic. A thick sedimentary complex containing greenstones and graywackes, etc. formed during the Palaeozoic and Mesozoic is found in the Intermontane Belt. Some of these have been subjected to metamorphism, and metamorphic rocks belonging to the prehnite-pumpellyite facies are occasionally found. More highly metamorphosed rocks belonging to the greenschist to amphibolite facies are distributed throughout the adjoining Omineca Crystalline Belt (Monger and Hutchinson, 1971*). The Rocky Mountain Belt on the eastern margin is an extremely mountainous area with a topographic relief in excess of 2500 m, and forms the continental divide of North America. An almost linear valley, the '*Rocky Mountain Trench*', runs between the Rocky Mountain Belt and the Omineca Crystalline Belt for about 1600 km from a latitude of 59° to 47 °N. This valley is remarkably narrow relative to its length, being about 30 km across in its wider sections and only a few kilometres in its narrower areas, and as its name implies, takes the form of a deep gorge (1000–2000 m deep). The Rocky Mountain Trench is said to have been formed before the Mesozoic or by the Cretaceous, but there are no commonly accepted theories about this. It could perhaps be a graben zone associated with a huge right-lateral fault (King, 1969*).

The formation of the above geological zones in the Canadian Cordillera took place from the Mesozoic to the Cenozoic. The low angle thrust faults characteristic of the Canadian Rockies were certainly developed at about this time. The structures have been formed in close association with a series of movements which occurred in the western region of the North American continent (Monger *et al.*, 1972*).

(b)   The McConnell Thrust and the structure of the Canadian Rockies foothills

McConnell first discovered a thrust fault around Mount Yamnuska in the foothills of the Canadian Rockies. Mount Yamnuska is about 2100 m above sea level, a small mountain rising some 750 m above the river which flows along its foot. The Cretaceous shale outcrops from an altitude of about 1750 m downwards. The strata undulate considerably, but are almost horizontal overall with virtually no structural disturbance. The rest of Mount Yamnuska is composed of Cambrian limestones, also distributed almost horizontally. Irregular fissures filled up with calcite, are found sporadically, but no signs of major disturbances. The McConnell Thrust passes horizontally between these limestones and shales on Mount Yamnuska, and is inclined at an angle of about 45° to the west on its western edge (Figure 2.IV.2). Following the fault northwards, the fault plane can be seen to continue almost horizontally for about 6 km, and the thrust plane itself waves gently and appears as a synclinal structure. The McConnell Thrust is observed between strata of different geological ages, the fault plane dipping in general towards the west, and it has been identified widely throughout the Canadian Rockies, in places with an extremely low-angle thrust plane. The fault can be traced overland for 260 km north of Yamnuska and 140 km to the south, making a total length of about 400 km.

The area around the Canadian Rockies has long been known to be rich in mineral fuel resources such as coal, oil and natural gas (Figure 2.IV.3).Coal intercalated in Jurassic–Cretaceous strata has been excavated in the Canmore coalfield near Banff since 1886. Test drilling for oil was undertaken in southern Alberta in 1891. In 1936, it was established that oil could be obtained from a depth of about 2000 m in Turner Valley. As a result of this

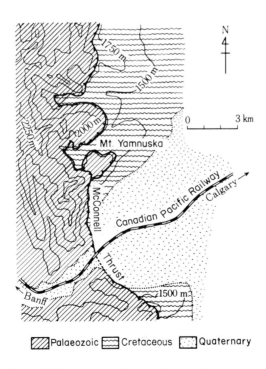

☑ Palaeozoic ❖ Cretaceous ⬚ Quaternary

Figure 2.IV.2 McConnell Thrust near Mount Yamnuska between Banff-Calgary in Alberta (after Clark, 1949* and Price and Mountjoy, 1970*)

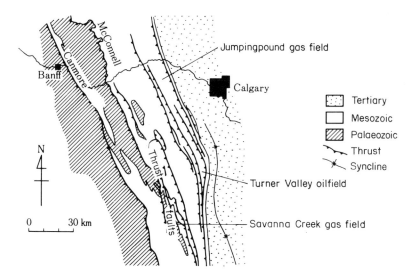

Figure 2.IV.3   Geological structures and the distribution of coal, gas and oil fields in the foothills of the Canadian Rockies (after Fox, 1959*)

discovery much more thorough surveys of the resources in this area were carried out, and this gradually led to the exploitation of natural gas and oil deposits in Jumpingpound, Turner Valley, Savanna Creek and so on. In Turner Valley the number of test drillings to date exceeds 500. These, together with surface survey work, have provided much information about the stratigraphy and have permitted detailed discussion about the under-ground structure. As shown in Figure 2.IV.4, the area may be divided into three parts by the westward-dipping Outwest Thrust and Turner Valley Thrust. Each block has been pushed up from the southwest to the northeast, superimposing older strata on younger. Except for local displacements due to folding or faulting, however, the order of superposition of beds within each block is normal, and no marked reverse structures are found. The block bounded by the Outwest and Turner Valley thrusts is 2000–3000 m thick; but the distance over which each fault extends on the map is more than 160 km in the former case, and about 70 km in the latter. Thus the block cut by the two thrust faults is quite thin in relation to its spread. In shape it resembles a sheet, and the overthrust block hence derives its name of '*thrust sheet*'.

(c)   Shape of thrust planes and associated folded structures

There are many thrust planes in the Canadian Rockies that are inclined at relatively steep angles of 40 – 60°, near the surface. The inclination tends to

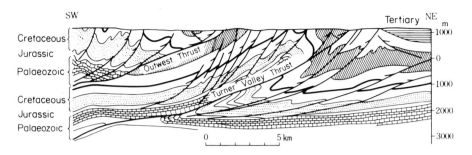

Figure 2.IV.4   Subsurface structure of Turner Valley oilfield in Alberta (after Fox, 1959*)

lessen gradually the deeper down one goes. The geological profile shown in Figure 2.IV.4 clearly demonstrates that these thrust planes are curved inwards, the whole being of a listric† shape as if gouged out with a scoop. Moreover several thrust faults perceived separately on the surface may converge into a single fault deeper down. This form of faulting is a feature frequently encountered in the area.

If the thrust plane is traced over a wide area, in some places it is relatively steep and in others much more gentle. According to Dahlstrom (1970*), who surveyed the morphological types of thrust faulting in the Canadian Rockies, when a thrust fault passes through a relatively competent layer as compared with those above and below, the fault will cross the bedding plane at a steep angle. Conversely, if it passes through a relatively incompetent layer as compared with the strata above and below, the fault plane will run almost parallel to the bedding plane. Figure 2.IV.5 demonstrates these relationships by restoring each layer to the position it occupied prior to the thrust faulting.

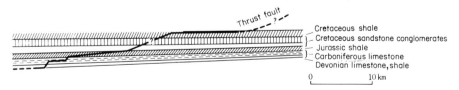

Figure 2.IV.5   Reconstructed thrust fault. The relationship between fault plane and bedding plane varies considerably, depending on the properties of the rock. When traversing an incompetent layer, the fault runs almost parallel to the bedding plane (after Dahlstrom, 1970*)

†English term derived from 'listrische' (Fläche) used by E. Suess (1909) when discussing the low angle Glarus overthrust in the Swiss Alps.

Just how far can the thrust faults seen in certain places be traced on a map? Or, what are they like at their extremities? Elliot (1976*) selected an area of the Canadian Rockies for which detailed geological information was available, and conducted such an examination of the extremities of thrust faults. There are limits to the length of a thrust fault on a map. If one looks along the direction in which it extends, the amount of displacement in the faults declines towards its two ends until it finally becomes impossible to trace further. Folded structures then appear instead, from the ends of the fault towards what would be its extensions. In other words, a structure showing the discontinuous morphological pattern of a fault turns into one with the continuous morphological patterns of folding. The area over which these folded structures are manifested appears to be about 8 km in the direction of extension of the fold axis. This distance hardly appears to be related to the scale of the main thrust fault itself. The fold axis generally appears to run about 0.5 km sideways from the end of the thrust fault, i.e. from where discontinuous forms of structure cease to be identifiable. The examples shown in Figure 2.IV.6 illustrate these relative positions.

Careful examination of a thrust fault reveals it to trace out a bow-shaped curve on the map (Figure 2.IV.7). If the bend of the fault line is assumed to depend on differences in the horizontal component of displacement due to the fault, then this amount will be zero at each end and reach a maximum in the central part of the curve. When one examines the direction of displacement in the said thrust fault, it will intersect a straight line joining the two ends of the bow-shaped curve at right angles. Elliot (1976*) explained that a relationship such as shown in Figure 2.IV.8 exists between the length (l km) of the fault on the map and the amount of displacement (u km) in its central

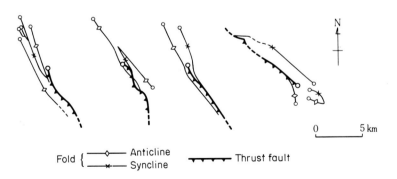

Figure 2.IV.6   Folded structures seen at the extremities of thrust faults. ○ shows end of each individual structure

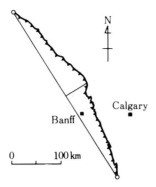

Figure 2.IV.7   Configuration of McConnell Thrust on a map (Elliot, 1976*)

part in the case of the Canadian Rockies thrust faults. In other words, he took $u = kl$, and $k = 0.07$. The displacement of approximately 40 km in the central part of the McConnell Thrust is judged to have taken about 8 million years to be produced, because the Cretaceous strata (laid down about 72 million years ago) were deformed by this thrust fault, and the structures formed by these movements were then covered by the Palaeogene formations (laid down 60–64 million years ago). Assuming displacement to have proceeded continuously, it therefore averages out at $du/dt = 4 \times 10^6$ cm/8×10^6 years = 0.5 cm/year. From the relationship shown in Figure 2.IV.8, du/dt 0.07 dl/dt, i.e., $dl/dt = 7$ cm/year. The ends of thrust faults turn into fold structures. The nearer the fold structures are to the ends of the faults, the more their form

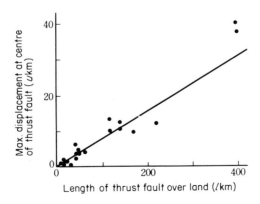

Figure 2.IV.8   Length of fault on map and maximum displacement in its central part
(Elliot, 1976*)

changes from open to closed. As the degree of deformation increases so they develop into faults. This suggests that a thrust fault propagates laterally at the average rate of 7 cm/year by the formation of fold structures at its ends.

Displacement due to a thrust fault decreasing towards the two ends produces secondary structures changing along the fault; in the central part of the fault where displacement is greatest, folding may develop parallel to the trend of the fault. The fold axis plunges towards both ends, with a *culmination* in the centre and depressions at the terminations. The relationship between thrust faults in the Canadian Rockies and their accompanying plunging folds is shown in Figure 2.IV.9.

(d) Formation of the Canadian Rockies

Figure 2.IV.10 shows a geological map of the area around Banff in Alberta. The Canadian Rockies are built up by incorporating numerous thrust sheets

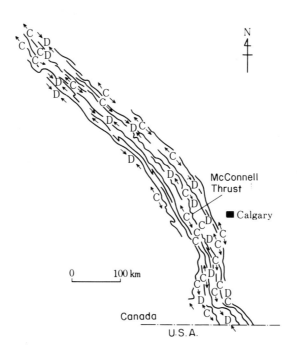

Figure 2.IV.9   Thrust faulting and fold structures in the Canadian Rockies. Solid lines indicate faults, arrows the plunge of fold axis. When the fold axis plunges in both directions, a culmination (C) of the axis results, and if the plunges of the axes converge a depression results (Dahlstrom, 1970*)

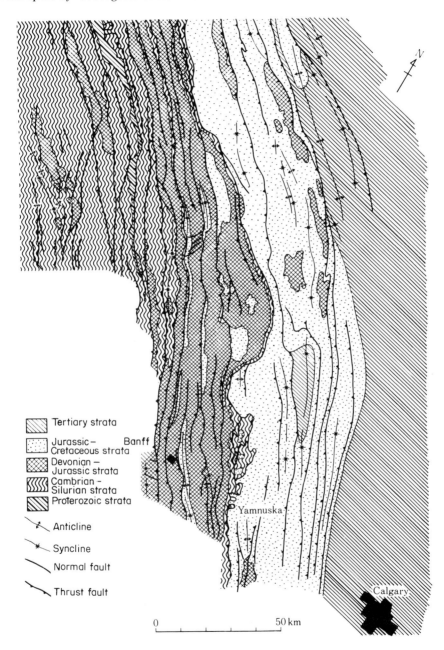

Figure 2.IV.10 Geological map of the Banff area, Alberta (after Wheeler *et al.*, 1972*)

with low angle, parallel thrust faults typified by the McConnell Thrust. The structure inside each thrust sheet is disturbed locally by drag folding and secondary faulting, formed in the wake of the thrust movements. However, no matter what the scale of the faults and folds, they are always arranged parallel to the mountain system, implying that they were formed in a single series of movements at almost the same time.

Surface surveys have clearly established that the majority of thrust faults slope steeply near the surface, but even now we are still uncertain just how deep their effects penetrate into the earth. Results obtained by seismic surveys and the analysis of seismic waves all over the Canadian Rockies have led to the belief that the boundary between the crystalline basement rocks, consisting mainly of Archaean metamorphic rocks and Cambrian strata declines gently in a uniform fashion to the west, and that these basement rocks do not participate in the thrust movements. In other words, there is a major plane of discontinuity (*decollément*) between these basement rocks and their sedimentary cover (or within the sedimentary strata), and this forms a tectonic boundary between the considerably displaced and deformed upper strata and unaffected lower section. It could be that all the major thrust faults, including the McConnell Thrust, converge and are linked at this plane of discontinuity.

In spite of the fact that upper and lower blocks have moved considerable distances in opposite directions along the plane of the thrust fault, it is not usual for large sheared zone to be encountered. In an example described by O'Brien (1960*) the fault runs through Cambrian olive-green chloritic shale. This shale is thought to act as a lubricant for the movements due to the faulting. Field studies and drilling samples indicate that when a fault develops, following the bedding of a layer, as described earlier, the layer is characteristically more incompetent than those above and below and internal slippage readily occurs within it. At the same time, the permeability of this layer to water is low. Water contained in it or in the layer above is unable to flow out freely, and hence may sometimes build up to considerable pressures. Such pore water finally builds up to such a high pressure that it becomes able to support the weight of the whole rock formation. This in itself can act as a lubricant and explains why large scale gliding takes place even on a slightly inclined surface (Hubbert and Rubey, 1959*). The fact that strata with such properties as those described above are observed along faults in the Canadian Rockies suggests that thrust faulting developed as a result of the gravity sliding of the rock mass on a major plane of discontinuity.

If the stratigraphic sequence and form of the thrust fault are understood, it is possible to estimate the relative displacement on both sides of the fault with some precision. Price and Mountjoy (1970*) measured displacement across the Canadian Rockies as a whole, on the basis of which they attempted to

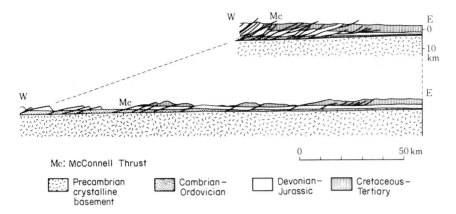

Mc: McConnell Thrust

Precambrian
crystalline
basement

Cambrian–
Ordovician

Devonian–
Jurassic

Cretaceous–
Tertiary

Figure 2.IV.11   Reconstruction of geological structures in Canadian Rockies (after
Price and Mountjoy, 1970*)

reconstruct the Rocky Mountain Belt (Figure 2.IV.11). As a result they
deduced that shortening due to the thrust fault groups would be in excess of
200 km. When one remembers that the distance from Calgary to the *Rocky
Mountain Trench*, which covers the whole of the Canadian Rockies, is about
160 km, the amount of shortening would seem incredible. However, this does
not represent contraction of the crust, because the crystalline basement rocks
in Figure 2.IV.11 are not believed to take part in displacement and deformation by thrust faulting. In summary, low angle thrust faulting in the Canadian
Rockies developed in the late Cretaceous, perhaps before the creation of the
Rocky Mountain Trench, as a result of the eastward sliding of superficial
sedimentary rocks deposited from the Cambrian to the Cretaceous time.
Pebbles derived from the Omineca Crystalline Belt, and transported to the
northeast are found in the upper Jurassic conglomerates of the Canadian
Rockies. The basement of the Canadian Rockies, therefore, probably began
to rise from about that time onwards. Sedimentary formations uplifted in
these movements began to slide northeastwards due to the increased potential
energy. Not only the features of the deformation structures but also their
distribution and the properties of the sedimentary rocks, have led to the
conclusion that the movements which created these geological structures
commenced during the late Jurassic and ended during the Eocene–Oligocene.
However, the periods of vigorous movement were probably between the late
Jurassic and the early Cretaceous in the west, and during the late Cretaceous
in the east.
    There are no marked angular unconformities in the sedimentary sequence
from the Cambrian to the Cretaceous in the Canadian Rockies, nor any

remarkable disturbances during that time. From the Cambrian to the Jurassic, the region had been relatively stable, but it was suddenly subjected to a great movement in the Cretaceous. Such changes were probably related to the development of the Omineca Crystalline Belt. It is conceivable that the rise of the core of the Canadian Cordillera triggered the gravity sliding and caused the low angle thrust faulting in the Canadian Rockies. These wide-ranging, huge structural movements can also be identified throughout Wyoming, Idaho and Utah in the USA, and are referred to there as *Laramide orogeny*.

Shinjiro Mizutani

References

Clark, L. M. (1949): Geology of Rocky Mountain front ranges near Bow River, Alberta, *Am. Assoc. Petrol. Geol. Bull.*, **33**, 614–33.

Dahlstrom, C. D. A. (1970): Structural geology in the eastern margin of the Canadian Rocky Mountains, *Can. Petrol. Geol. Bull.*, **18**, 332–406.

Elliot, D. (1976): The energy balance and deformation mechanisms of thrust sheets, *Phil. Trans. Roy. Soc. London*, A 283, 289–312.

Fox, F. G. (1959): Structure and accumulation of hydrocarbons in southern foothills, Alberta, Canada, *Am. Assoc. Petrol. Geol. Bull.*, **43**, 992–1025.

Hubbert, M. K. and Rubey, W. W. (1959): Role of fluid pressure in mechanics of over-thrust faulting. I Mechanics of fluid-filled porous solids and its application to over-thrust faulting, *Geol. Soc. Am. Bull.*, 70, 115–66.

King, P. B. (1969): Tectonic Map of North America, 1:5 000 000, U. S. Geol. Survey.

McConnell, R. G. (1887): Report on the geological structure of a portion of the Rocky Mountains, accompanied by a section measured near the 51st parallel, *Geol. Survey Canada, Ann. Rept., New Ser.*, 2, 5D–41D.

Monger, J. W. H. and Hutchinson, W. W. (1971): Metamorphic map of the Canadian Cordillera, *Geol. Survey Canada, Paper* 70–33, 71 pp.

Monger, J.W.H., Souther, J.G. and Gabrielse, H. (1972): Evolution of the Canadian Cordillera: a plate-tectonic model, *Am. J. Sci.*, **272**, 577–602.

O'Brien, C.A.E. (1960): The structural geology of the Boule and Bosche Ranges in the Canadian Rocky Mountains, *Q. J. Geol. Soc. London*, **116**, 409–36.

Price, R. A. and Mountjoy, E. W. (1970): Geologic structure of the Canadian Rocky Mountains between Bow and Athabasca rivers — a progress report, *Geol. Assoc. Canada, Spec. Paper, No. 6*, 7–25.

Price, R. A. and Mountjoy, E. W. (1970): *MAP 1265 A*, Geology, Canmore (East Half), Alberta, 1:50 000, Geol. Survey Canada.

Wheeler, J. O., Campbell, R.B., Reesor, J. E. and Mountjoy, E. W. (1972): *Structural style of the southern Canadian Cordillera*, Excursion A–01–X–01 Guidebook, XXIV IGC, 1972 Montreal, Quebec, 118 pp.

V   *Nappe structures in the Swiss Alps*

The town of Glarus is located some tens of kilometres southeast of Zurich, the largest city in Switzerland. This area is one of the oldest provinces in Switzerland running along the Linth gorge. In the mid-eighteenth century it was the heart of the Swiss cottage textile industry, and after the beginning of the nineteenth century developed into a major centre for Alpine mountaineering. At about this time the geologist Escher, a native of the area, made an important observation. The Linth ravine is a U-shaped valley gouged out by glaciers in the Würm–Riss period, and bare mountains tower 2000–3000 m high on both sides. Escher noticed that although volcanic conglomerates of the Permian system, known as Verrucano, are distributed throughout the ravine near the mountain tops, Eocene strata composed of sandstone and mudstone, known in Switzerland as 'flysch', can always be found lower down. Lyell published his *Principles of Geology* at about this time, and it began to be accepted scientific fact that 'lower strata are older than upper strata'. Escher, who was not only a Swiss geologist but also an excellent 'Alpinist', is said to have made sceptical colleagues verify his major observation with their own eyes by taking them around the rock faces. Soon no one could doubt that this stratigraphical inversion continued for a distance of at least 20 – 30 km. Escher made a painstaking survey of the Linth ravine and forced people to see that amazing crustal movements had occurred at that period which no one would have expected. In short, Verrucano, which had initially been deposited at a lower level, had migrated over Tertiary strata, as a result of almost horizontal movement along a fault.

Such large scale faulting came to be known as *overthrusting* and it has been realized that the high mountain peaks of the Swiss *Alps* had been formed mainly by this process. A large body of rock that has migrated for a considerable distance by such overthrusting is known as a 'nappe'. When it was finally appreciated between the late nineteenth and early twentieth centuries that overthrusts are common structural features, encountered in high mountainous areas, the Swiss Alps had come to be described in most textbooks as a model of geological structure caused by orogeny.

The fault discovered by Escher between the Verrucano and flysch was called the *Glarus Overthrust*. There is a good outcrop of the fault by the road at Schwanden, a small town about 5 km upstream from Glarus (Figure 2.V.2). Nearby, on a metal plate, is inscribed 'overthrust faults were first discovered here on 1st August 1840 by Arnold Escher von der Linth, laying the foundations of Alpine geology'.

Studies of the Swiss Alpine structures were subsequently pursued by Albert Heim, a disciple of Escher, and his results are summarized in his two volume

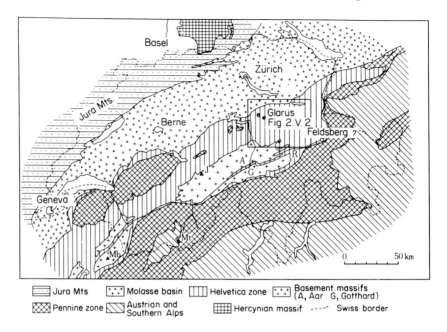

Figure 2.V.1   Generalized structural map of Switzerland. J — Jungfrau (4158 m); Mb — Mont Blanc (4807 m); Mt — Matterhorn (4477 m); R — Upper reaches of Rhine

Geologie der Schweiz (1919–22). This work is, even now, one of the most important pieces of literature on alpine geology, and the major structural classification is still used in its original form.

The structural features of the Swiss Alps may be divided into molasse basins, the Helvetic zone, Pennine zone and Austrian and Southern Alpine zones (Figure 2.V.1). The most important parts are the Helvetic and Pennine zones which both contain many of the structural features related to overthrust faulting. Most parts of the Pennine zone have been subjected to later metamorphism and consist of more complex structures so that it is difficult to trace the early deformation history. We, therefore, concentrate below on the Helvetic zone and particularly on the area around Glarus, which is representative of Alpine structures and has also been extensively studied.

(a)   The Glarus Overthrust and Helvetic nappes

The area shown in Figure 2.V.2 lies about 50 km east of the renowned Jungfrau (4158 m). In this area, the mountains consist of 3000 m peaks permanently covered with snow, the highest being Tödi (3620 m). In the

lower part of the diagram are the central massifs, which were formed as a
result of Hercynian orogenic movements and are composed of metamorphic
rocks and granites. The Helvetic zone consists of Permian to Palaeogene
formations deposited on these Hercynian basement rocks.

The main part of the Permian formation is a characteristic volcanic
conglomerate called Verrucano, which reaches thicknesses of about 1000 m in
the vicinity of Glarus. The Triassic system is much thinner and includes red
slate. In terms of thickness, the Helvetic zone is mostly composed of Jurassic
and Cretaceous carbonate rocks. The Jurassic system is about 1000 m thick,
and particularly in its upper part, is composed principally of fine grained

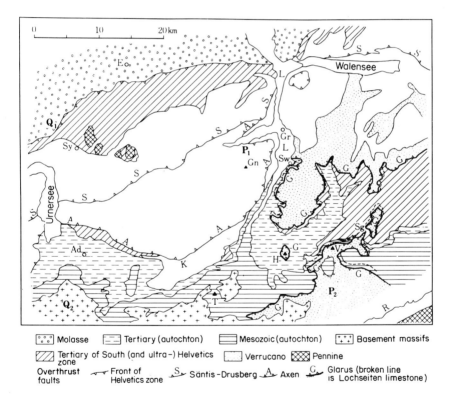

Figure 2.V.2   Tectonic map of the eastern Helvetic belt (after Tektonische Karte Der
Schweiz 1/500 000. Schweizerischen Geologischen Kommission 1972* partly revised by
the author). Ad — Altdorf; E — Einsiedeln; Gn — Glärnisch (2914 m); Gr — Glarus;
H — Hausstock (3158 m); K — Klausen Pass; L — Linth Valley; Sg — Segres Pass; Sw
— Schwanden; Sy — Schwyz; T — Tödi (3620 m); V — Vorab (3028 m); P_1, P_2, Q_1,
Q_2: cross-section — see Figure 2.V.5

limestones. The later Jurassic limestone is widely distributed from the Swiss Alps to the Jura mountains (on the Franco-Swiss border) and Bavaria in southern Germany. In the Helvetic zone the limestone is black and partly muddy, but further north it changes to a greyish reef limestone. At the beginning of the Cretaceous, it seems that an area further south of the present position of the Aar massif (Figure 2.V.1) became a centre of sedimentation. The lower Cretaceous system is no more than 100 m thick to the north of the Aar massif, but it increases in thickness, up to about 1500 m in the south. The mid-Cretaceous is represented by a 10–80 m thick, glauconite-bearing formation known as 'gault' which forms a convenient marker bed when we take a look at deformation of strata from a distance.

The Glarus Overthrust came into existence when strata and rocks on top of the Permian and Cretaceous separated off and moved northward from the Aar and Gotthard massifs. The nappe migrated for a distance of at least 40 km to the north, from the upper Rhine, over the Aar massif, slipping over Jurassic and Cretaceous formations and even Tertiary formations deposited on these Mesozoic rocks. Looking at the height of the present fault plane, we find that it goes through the high peaks such as Hausstock (3158 m) almost horizontally at an altitude of about 2800–3000 m (Figure 2.V.3). However,

Figure 2.V.3   Glarus overthrust in Segres Pass (after Heim, 1921*) 1 — Verrucano; 2 — Lochseiten limestone; 3 — flysch

because the altitude declines from here to about 700 m near Schwanden, about 14 km to the north, the fault plane has an average inclination of 8° north. Most of the Glarus nappe currently consists of Verrucano only, as can be seen from Figure 2.V.2; in the northern part of the nappe, however, there remain thick Jurassic to Palaeogene strata as well as Verrucano, and so the Glarus Overthrust must have moved with an over-burden of at least 2000 m.

Thus the nappe transported by the Glarus overthrust consisted of the sedimentary rocks originally deposited in the geosyncline in the southern part of the Aar massif. The strata deposited in the north of the Aar massif virtually remained in their original position and were covered by this nappe. Such a block exhibiting no migration is called an '*autochton*', while as mentioned before, the block exhibiting migration with overthrust is called 'nappe'.

We have already discussed the stratigraphy of the area as far as the middle Cretaceous. Towards the upper Cretaceous the proportion of shale to limestone increased and the lithology changed gradually into a shale-sandstone facies. In Palaeogenic, rocks are mainly composed of marine clastic sediments, known as *flysch*. Flysch was the dialect term used originally among the people of central Switzerland for shaly rocks, which produced moist acid soils and caused superficial landslides, but the term also came to be used for Cretaceous and Palaeogene marine formations in Swiss Alps.

Important research on the Swiss Alps after the Second World War was carried on to investigate the sedimentology and paleogeography of the area prior to the development of overthrusting. Trümpy as a representative of the researchers summarized the facies variations in the flysch formations as shown in Figure 2.V.4 (Trümpy, 1960*). This figure assumes that the autochtonous belt and Helvetic nappe were situated on the north and the south side of the Aar massif, respectively. The ultra-Helvetic zone refers to the nappe that moved from even further south than the original position from which the *Helvetic nappe* was derived. We are unable today to fully understand the picture as a whole, because of erosion, weathering and sedimentary covers, but the further south one goes, the older the flysch formations appear to be.

In certain places, upper Cretaceous and Palaeogene formations can be found between the Glarus overthrust and the flysch. These are the nappes which migrated from the south as a result of separate thrust faults older than the Glarus overthrust, and contain both Helvetic and ultra-Helvetic nappes. The area to the south of the present Aar massif was uplifted prior to the movement of the Glarus fault by these previous overthrust faults. Consequently Cretaceous to Eocene flysch, which was then the uppermost part of the sedimentary pile, slid over the autochton on the northern side probably in Oligocene.

It is rather difficult to observe the Glarus overthrust in the area west of the Linth valley. Typical structures encountered there include the Axen and

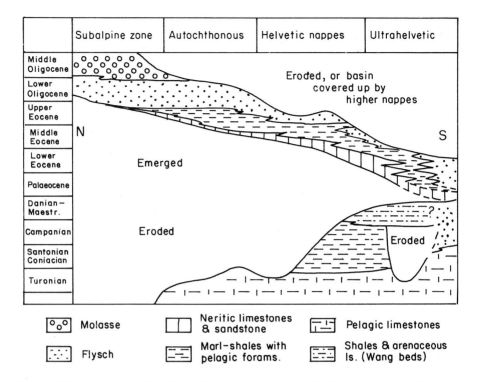

	Subalpine zone	Autochthonous	Helvetic nappes	Ultrahelvetic
Middle Oligocene				
Lower Oligocene				
Upper Eocene				
Middle Eocene				
Lower Eocene				
Palaeocene				
Danian–Maestr.				
Campanian				
Santonian Coniacian				
Turonian				

Figure 2.V.4   Age relationship of flysch formations in the Helvetic zone of Eastern Switzerland

Säntis-Drusberg nappes. The base of the Axen nappe is Triassic, but near Lake Urner (left in Figure 2.V.2) the middle Jurassic or the lower Cretaceous formations are laid directly on the flysch. On the other hand, in the Säntis-Drusberg nappe lower Cretacean formations, constituting the base of the nappe are directly laid over the Axen nappe. The Axen thrust is found to be higher than the Glarus nappe along the Linth valley and thus, it is certainly later than the Glarus thrust. The Säntis-Drusberg thrust on the northernmost part of the Helvetic belt is found there to be higher than the Axen thrust. Generally speaking, these three major overthrust faults moved northwards in succession. They reached the position of a molasse formation, finally stopped, and were pushed back by the molasse body.

Coarse clastic rocks containing extremely large boulders are distributed throughout the north of the Helvetian zone. This is the so-called *molasse* whose development spanned from the middle Oligocene to the Pliocene. As the pebbles are all derived from the Helvetic nappe, the molasse is believed to have formed naturally from the detritus which had fallen down from the mountains of the Helvetic nappe. Two sedimentary cycles are seen in this molasse; the earlier one involved a change from marine to continental deposits during the Oligocene, and the later was again a change from marine to continental in the Miocene.

The cross-section $Q_1 - Q_2$ in Figure 2.V.5 illustrates the general structure of the Axen and Sätis-Drusberg nappes seen along Urnersee. Unlike the Glarus nappe, which is shown on $P_1 - P_2$ section of Figure 2.V.5, the structure of $Q_1 - Q_2$ is more complicated. This is because they exhibit a backward-moving structure at the very end of the thrust faulting. By tracing marker beds one can deduce that both were probably once one nappe, which later split. Although the nappe ceased to be able to move when it encountered a resistant body such as the molasse, the upper layers continued to move still further, so forming complex folding by the lateral turning or inversion of the anticline as illustrated in the figure. The term '*recumbent fold*' was proposed for this type of folded structure. The main structural framework of Helvetides has been completed through the movement of the Glarus, Axen and Säntis-Drusberg overthrust movements. The time of this movement is thought to have corresponded to the second cycle of molasse in Miocene.

(b) The formation of nappe structures

Although the Helvetic zone is about 50 km wide and extends for a length of 200 – 300 km from Switzerland to the Mediterranean, its structural pattern does not differ fundamentally from those around Glarus. The folded struc-

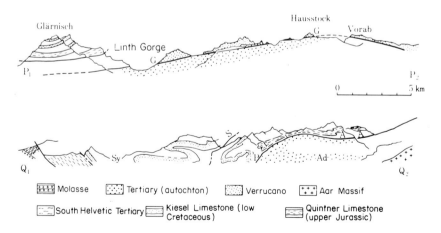

Figure 2.V.5   Structural profile of Helvetic zone. A — Axen overthrust; G — Glarus overthrust; S — Säntis-Drusberg overthrust; P — see Figure 2.V.6; Ad, Sy, P_1, P_2, Q_1, Q_2 — see Figure 2.V.2

tures of the Pennine belt are considered essentially the same as those in the Helvetian zone. What kind of mechanism could have formed the overthrust faults seen in these areas? Places where one may gain a direct understanding of thrust planes are limited even in the Alps, where towering rock faces proliferate. Schwanden in *Glarus*, mentioned above, is an exceptional locality and the Glarus overthrust observed there has been described in detail by Heim (1921–1922). The Verrucano and flysch are normally separated by a 1 –3 m thick transitional band, which is composed of a fine-grained limestone, containing numerous shale-like patches several millimetres wide, and is known as Lochseiten limestone. It exhibits flow structures as can be seen in Figure 2.V.8, and is considered to have taken the role of a lubricating agent for the thrusting. However, the fault itself is a single plane running through the Lochseiten limestone. The origins of Lochseiten limestone are still uncertain, but lithologically it is quite similar to limestones of the upper Jurassic system in the eastern Swiss Alps.

The sequence of fault movements is believed to have been as follows: Displacement along the Glarus overthrust accelerated rapidly from a movement when, for whatever reason, the Lochseiten limestone became involved in thrusting. The Lochseiten limestone then began to flow and behaved as a lubricant, and major movements occurred forming the latest fault plane. Verrucano is a fine-grained and very hard, compact rock. In the Glarus fault movements, a rock body of 1000 – 2000 m thick including Verrucano and the overlying Mesozoic formations slid over the Lochseiten limestone in virtually its original form without any internal deformation.

Figure 2.V.6 shows the deformation of the lower Cretaceous limestone in the centre of the Axen nappe on the Axen-road by Lake Urner. This is an example of the most extreme deformation in the Helvetian zone. The limestone layers, 3 – 5 m thick, have formed small scale recumbent folds. It may reasonably be said that in this case the model of deformation is flexural slip folding because little variation in the thickness of each layer can be observed, particularly in the hinge zones, and no elements indicative of flow movement are apparent, and because the recumbent anticlinal section is a bend due to fracturing. Other geological considerations, including the fact that even the lowest sedimentary formations of both the nappe and autochton are not buried deeper than a few kilometres, and that the strata have not been affected by extremely high temperatures and pressures to produce metamorphic rocks, have led to the conclusion that these folds must have been produced at a shallow depth in the crust.

The dynamic mechanism whereby a 2 km thickness of rock could slide horizontally over 30 – 40 km, as was found at Glarus has been studied in many ways since the beginning of the twentieth century. However, the early conclusions were that it was physically impossible to produce this type of movement with a simple model of horizontal friction.

Figure 2.V.6 Folds of lower Cretaceous limestone in Axen nappe (locality P in Figure 2.V.5)

Hubbert and Rubey (1959*) applied the ideas of rock mechanics to overthrusts like Glarus. If one considers a nappe of thickness Z and length x as in Figure 2.V.7, and take μ_1 as the coefficient of friction between the nappe and autochton, or ϕ as the angle of friction, then the shearing stress τ_{zx} and vertical stress σ_{zz} acting on the nappe are as follows:

$$\tau_{zx} = \mu_1\sigma_{zz} = \tan \varphi\sigma_{zz}$$

and

$$\sigma_{zz} = \rho g z \qquad (2.\text{V}.1)$$

ρ and g are the density of the nappe and gravitational acceleration respectively. The lateral pressure σ_{xx} acting on the nappe is dependent on Coulomb's failure criteria (Chapter 3, p. 140):

$$\sigma_{xx} = a + b\sigma_{zz} \qquad (2.\text{V}.2)$$

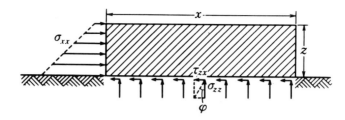

Figure 2.V.7   Hubbert's model for overthrust faulting (after Hubbert and Rubey, 1959*)

In this equation a and b are coefficients obtained from cohesion inside the nappe $\tau0$ and the coefficient of internal friction μ_2 using the following equations:

$$a = 2\tau_0 \sqrt{b}$$

$$b = \frac{\sqrt{1+(\mu)^2} + \mu_2}{\sqrt{1+(\mu)^2} - \mu_2}$$

As the lateral pressure and friction are equal at the moment the nappe starts to move

$$\int_0^z \sigma_{xx} \, dz - \int_0^x \tau_{xx} \, dx = 0 \qquad (2.V.3)$$

Substituting equations (2.V.1) and (2.V.2) and solving equation (2.V.3) gives the following conditions of sliding for the nappe

$$x = \frac{a}{\rho g \mu_1} + \frac{b}{2\mu_1} z \qquad (2.V.4)$$

However, if one substitutes values such as $\mu_1 = \mu_2 = 0.577$, and $\tau_0 = 200$ kg/cm² into this equation and solves it, one obtains length of the nappe of $x = 10.6$ km for a given thickness of 2 km, and thus we cannot explain horizontal faulting on the scale of Glarus. Hubbert and Rubey then turned their attention to the pressure of interstitial pore water as a means of solving the problem. Pore water in rocks is normally at a pressure equal to that of hydrostatic pressure at that depth. At about that time the experiments of Handin *et al.* made it clear that the effects of *effective confining pressure*, a well-known concept in soil mechanics, are also applicable to solid rocks. Generally speaking the mechanical properties of rock, such as its strength, will vary markedly with the pressure acting around the rock, i.e., the confining pressure (or in this case overburden pressure of the nappe). In other words, the mechanical properties of the rock are dependent on the confining pressure. The effective confining pressure defined as confining pressure minus pore water pressure is what actually determines the mechanical properties. Therefore, if the interstitial pore water pressure increases, everything else remaining equal, the effective confining pressure must decrease, as also do the strength and coefficient of friction. In fact, at about the same time in the oilfields of Texas it was found that deeply buried sediments commonly possess abnormally high pressures in which the interstitial pore water pressure was higher than the hydrostatic pressure. Hubbert and Rubey demonstrated that thrust faults 50–100 km long could easily arise from changes in such mechanical properties by revising their calculations to include such abnormal high pressure conditions in equation (2.V.4). The work of Hubbert and Rubey was quite revolutionary in providing an initial answer to a problem outstanding for half a century, but it would be open to criticism on several points, as is the majority of such leading research. First of all there are physical problems as to whether their equations and models are reasonable. It is doubtful whether the abnormally high pressures required could have been generated in the layers of very fine-grained and compact rock with small porosity as are seen around Glarus. To this question Hubbert and Rubey replied that the Texas oilfield is currently in a geological condition which may be regarded as a geosyncline and that abnormally high pressures could be generated quite commonly during sedimentation in a geosyncline.

Whether or not abnormally high pressure is common in orogenic belts uplifted from geosynclines still remains a problem, and another outstanding problem is the role played by Lochseiten limestone during movement on the Glarus overthrust. It was clearly established, from the time of Heim onwards, that the Lochseiten limestone could be found with virtual certainty along the whole of the Glarus fault plane (Figure 2.V.2) and Heim soon realized that it had acted as a lubricant between the hard Verrucano and the flysch. The present author stayed in Zurich during 1970–72 and has had many opportunities to visit Glarus. His impression was that even the Lochseiten limestone could not have been the cause of the Glarus nappe sliding 40 km producing the largest horizontal fault in the Helvetic belt and perhaps in the world. The author conducted a series of experiments in which pressures of up to 3000 kg/cm^2 were applied to all the rock types found in eastern Switzerland, in order to examine their mechanical properties under high pressure conditions. The results showed that the upper Jurassic and Lower Cretaceous limestones were the weakest and most ductile (Hoshino, 1973). These characteristics were particularly marked with samples taken from the upper Jurassic system near the Aar massif. This limestone, known as Felsberg limestone, after the name of the place where it was collected, closely resembled the Lochseiten

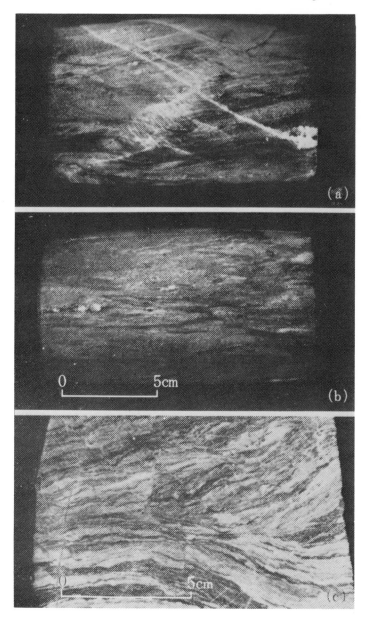

Figure 2.V.8   Felsberg limestone deformed under confining pressure of 500 kg/cm^2, at temperatures of (a) 20° and (b) 100°C, respectively. (c) Lochseiten limestone beneath the fault plane of Glarus overthrust

limestone in the sense that it also contained a banded zone, 0.5–2.0 mm wide, perhaps suggesting mudstone origins. Like most rocks, Felsberg limestone is brittle at room temperatures and forms shear fractures when it fails. At 100°C, however (with 500 kg/cm^2 confining pressure), it becomes highly ductile and exhibits flow. Figure 2.V.8 shows samples of Felsberg limestone deformed in the laboratory (a and b) compared with natural Lochseiten limestone (c). The pattern of deformation achieved in (b) is quite similar to that of the Lochseiten limestone (c) (Hoshino, 1977). The basal parts of nappes such as the Axen and the Säntis-Drusberg always contain various kinds of Jurassic or Cretaceous limestone, and the Lochseiten limestone was derived from the most ductile limestone amongst these. When these facts are considered together with the above experimental results, the existence of ductile limestone capable of considerable flow movement may clearly be seen as a vital condition in the formation of large overthrust faults. This may also be supported by comparing the nappe structures of the Swiss Alps to those of the southern Tyrol and the Dolomites. The principal lithologies found in the Dolomites and the southern Tyrol areas are thick layers of Triassic dolomite. Dolomite is a carbonate rock like limestone, but is very different physically. Whereas limestone is extremely ductile, dolomite is very brittle. This could perhaps be the reason why nappe structures are not found in the Dolomites or southern Tyrol.

The majestic nappe structures seen in the Helvetian zone today may perhaps have been created as the composite result of all the various factors discussed above.

Some 100 years after the discovery of the Glarus fault by Escher, the Swiss Alps have been treated as a model region for structural geology. However, several problems concerning the origin and mechanism of the overthrust faults still remain. For example, where did the Helvetic nappe originate? According to Trümpy (1960*) the width of the Swiss Alps is estimated as 150 km, but would be about 300 km if restored to their original positions prior to deformation. Many people, at present, believe that the root or source of the Helvetic nappe lies between the Aar and the Gotthard massifs (Rutten, 1969*). The uplift of the whole Alpine block that occurred after the nappe formation was also a most important event in the later stages of Alpine orogeny, and the existence of some lateral faults in the Swiss Alps indicates that this late stage uplifting was not caused by a simple upward movement.

Kazuo Hoshino

References

Heim, A. (1919): *Geologie der Schweiz. I Molasseland und Juragebirge*, 704 pp., Tauchnitz, Leipniz.

Heim, A. (1921–22): *Geologie der Schweiz. II Die schweizer Alpen*, Tauchnitz, Leipniz, (1) pp. 1–476, (2) pp. 477–1118.
Hubbert, M. K. and Rubey, W. W. (1959): Role of fluid pressure in mechanics of overthrust faulting. I. Mechanics of fluid-filled porous solide and its application to overthrust faulting, *Geol. Soc. Am. Bull.*, **70**, 115–66.
Rutten, M. G. (1969): *The Geology of Western Europe*, Elsevier, 520 pp. Amsterdam.
Schweizerischen Geologischen Kommission (1972): *Tektonische Karte der Schweiz*, 1/500,000.
Schweizerischen Geologischen Kommission (1972): *Geologische Karte der Schweiz*, 1/500,000.
Trümpy, R. (1960): Paleotectonic evolution of the central and western Alps, *Geol. Soc. Am. Bull.*, **71**, 843–908.

VI   The *Franciscan formation* and the *tectonic mélange*

From early Mesozoic times to the present day the East Pacific Rise has not only provided an outlet for the immense heat, which emanates from inside the earth, but has also been a site for the production of oceanic crust. The oceanic crust to the east and west of the Rise, known as the *Farallon* and *Pacific plates*, has been growing very slowly in both an easterly and a westerly direction. The eastwards-growing Farallon plate has collided with and is being subducted beneath the American continental block (American plate), which is migrating westwards as a result of activity along the mid-Atlantic Ridge and expansion of the Atlantic Ocean. Strata ranging from 15 000 to 30 000 m thick are distributed throughout the mountain belt, which follows the west coast of North America. They are principally composed of greywacke, sandstones and shales, fairly large quantities of cherts and basaltic marine volcanic material, together with smaller quantities of conglomerates and limestones, and are late Jurassic to Cretaceous in age. In 1895 Andrew C. Lawson, a Professor at the University of California at Berkeley named the strata found around San Francisco Bay, the 'Franciscan Formation'. The distribution of the Franciscan Formation seen today has been somewhat complicated by the *San Andreas fault* (see Section 2.II) and its accompanying minor faults, as shown in Figure 2.VI.1.

If one looks at the area after restoring the 600 km or so right-lateral horizontal displacement, which has occurred along the San Andreas fault since the Cretaceous, the Franciscan Formation is seen as a long narrow band about 150 km wide extending for 3000 km in a north–south direction, as shown in Figure 2.VI.2 (Suppe, 1970*). This long geosyncline, which existed along the western edge of the North American continent from the Jurassic to the Cretaceous, was on a scale which could have joined the Izu-Mariana

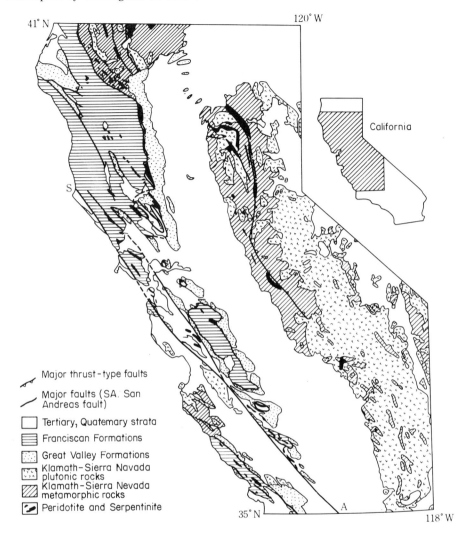

Figure 2.VI.1 Geological map of Franciscan zone (after Ernst, 1970*)

Trench and Kurile Trench to the Japan Trench. This was the place where the Farallon plate was being subducted under the eastern edge of the American plate (Atwater, 1970*).

The Mesozoic history of California which centres on the Franciscan Formation and the tectonic *mélange* provides us with perhaps the best example to demonstrate the kind of geological structures metamorphism created when an oceanic plate collides directly with a continental plate and the former underthrusts the latter.

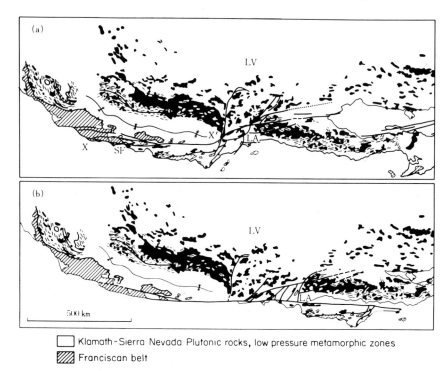

☐ Klamath‑Sierra Nevada Plutonic rocks, low pressure metamorphic zones
▨ Franciscan belt

Figure 2.VI.2   (a) Current distribution of the Franciscan zone. (b) Reconstruction of
the situation in the Cretaceous period produced by sliding the San Andreas fault 600
km back (Newport–Inglewood fault in southern California) (X to X′) (after Suppe,
1970*). In (b) both the Franciscan zone and the plutonic and low pressure metamorphic areas of the Klamath–Sierra Nevada zone become simple long belts running
parallel to each other. LA: Los Angeles; LV: Las Vegas; SF: San Francisco

(a)   The Franciscan Formation and metamorphism

Most of the sedimentary deposits in the Franciscan Formation, excluding
cherts and volcanic material, were supplied from the granitic and metamorphic rocks, which were exposed from the Klamath Mountains through the
Sierra Nevada ranges to the California Peninsula. Slump structures within the
sediments resulting from large-scale sliding from east to west of the ocean
floor deposits at the time of deposition can be observed in the Franciscan
Formation. Occasionally, graded bedding, ripple marks or flute casts are
found, and some sedimentary layers representing turbulent mud flows are
also present locally. Volcanic material accounts for some 10 per cent of the
Franciscan Formation, the majority being lava flows with pillow structure,

and agglomerate rocks with little fine-grained tuff. Single rock masses composed of lava-agglomerate rocks may reach thicknesses of up to 2000 m and lengths (i.e. horizontal diameter of the volcanic structure) of up to 10–16 km, their scale being similar to that of seamounts in the Pacific. Thin layers of tuff are often intercalated with sandstone, and thin layers of limestone or chert are also interbedded with the pillow lava. The Franciscan Formation generally becomes younger from the east to west, and no strata younger than late Cretaceous can be found in the east. Many 'cold' or 'Alpine' type peridotites and their associated gabbros, were intruded into the Franciscan Formation. Most of the peridotites are almost completely serpentinized and the gabbro has been severely affected by hydrothermal alteration.

An important feature of the Franciscan Formation is that it seems to have been deposited directly on the oceanic crust. Large quantities of serpentine are found within this Franciscan belt in particular on its eastern margin. Most of them are found along sheared parts of the Franciscan Formation. A considerable number of exotic rock masses of eclogite and metamorphosed basalts are also observed in the serpentinite and give a good indication of the type of basement which exists deep below the Franciscan Formation. If the basement were composed of sedimentary rocks, gneisses or granites of Mesozoic, Palaeozoic or Precambrian age one would expect to find some fragments of these rocks within the serpentinite masses, but no fragments of continental type crust have been found.

Until the publication of work by Bloxam in 1956, no one thought that rocks of the Franciscan Formation, in which schistosity is rarely developed and some beautiful fossils are preserved are metamorphosed rocks. The Formation was said to be ordinary sedimentary rocks with extensive faulting. However, in sandstone from the Franciscan Formation examined by Bloxam, the Calcic plagioclase and clays (chorite) in the original material had changed almost completely to lawsonite, jadeitic pyroxene, glaucophane, and quartz by the following chemical reaction (Bloxam, 1956*)

$$
\begin{array}{ccccc}
\text{Calcic plagioclase} & + & \text{chlorite} & + & \text{water} \\
(3CaAl_2 Si_2O_8 + 3NaAlSi_3O_8) & & (Mg_3Si_2O_5 (OH)_4) & & (5H_2O)
\end{array}
$$

$$
\begin{array}{cccccccc}
\text{Lawsonite} & & \text{Jadeite} & & \text{Glaucophane} & & \text{Quartz} \\
(3CaAl_2Si_2O_82H_2O) & + & (NaAlSi_2O_6) & + & (Na_2Mg_3Al_2Si_8O_{22}(OH)_2) & + & (SiO_2)
\end{array}
$$

Such sandstone layers were subsequently found to account for almost two-thirds of the Franciscan zone. It has long been thought that the blue-green fragments containing glaucophane were present as 'pebbles' in the sandstones forming the Franciscan Formation, and that glaucophane schist had already been formed before sedimentation of the Franciscan Formation occurred. However, these 'pebbles' of glaucophane schist are the recrystallization products of volcanic fragments in the sandstone with mineral assemblages including glaucophane due to the high-pressure regional metamorphism to which the Franciscan Formation has been subjected. The sandstone around the volcanic fragments has also been metamorphosed into mineral assemblages containing jadeitic pyroxene and lawsonite. The mystery surrounding the presence of glaucophane schist 'pebbles' in the Franciscan sandstone was the result of a misunderstanding that arose because of ignorance of how rocks with differing chemical compositions, i.e. feldspathic sandstone and basaltic volcanic rocks, recrystallize into different mineral assemblages under high pressure metamorphism.

It was, subsequently, found that most of the $CaCO_3$ minerals in the Franciscan Formation were not calcite but aragonite, and it is now clear that the whole of the Franciscan Formation has suffered regional metamorphism under conditions of a low geothermal gradient, i.e. 10–15 °C/km producing relatively low temperatures (150–300 °C) but abnormally high pressures (5–8 kb) (Ernst *et al.*, 1970*). Such high pressure metamorphism has been called '*glaucophane schist or blue schist regional metamorphism*'. Worldwide, glaucophane schist regional metamorphism is found in a belt within relatively young, Phanerozoic, orogenic belts such as in southern Chile, Cuba, the California Coast Ranges, Oregon, the west coast of Canada, southern Alaska, Kamchatka, Sakhalin, Hokkaido, Honshu, Shikoku, Kyushu, Taiwan, the Philippines, Celebes, New Guinea, New Caledonia, New Zealand, Turkey, southern Italy, Corsica and the Alps. Only a very few instances have been reported of glaucophanitic schists in relatively old metamorphic areas of Precambrian or Palaeozoic age, these include isolated occurrences in Wales, Scotland and the north coast of France, and eastern Siberia. The metamorphism to which the Franciscan Formation was subjected had the highest pressure/temperature ratio in the world. Other distinct characteristics of the metamorphic history in the Franciscan Formation are related to the geologic development of the tectonic *mélange* which we discuss later.

Isotopic examination has indicated that regional metamorphism in the Franciscan Formation occurred 70–150 Ma, i.e. in the late Jurassic–Cretaceous, when the strata were being deposited. It should be mentioned that the metamorphic period indicated by isotopic methods and the period of deposition of the Franciscan Formation deduced from fossils are strikingly similar. In other words, high pressure regional metamorphism of the Formation occurred at a depth of 20–30 km at the same time, in the Jurassic, as the geosynclinal sediments, up to 30 000 m thick were laid down, with violent marine, basaltic, volcanic activity and accompanied by rapid subsidence at about 5–10 mm/1000 years (Suppe and Armstrong, 1972*; Suppe 1972*).

(b) The tectonic mélange

Another important feature of the Franciscan Formation is the presence of coarse-grained metamorphic rocks, showing relatively high grades of metamorphism, reaching epidote amphibolite facies or eclogite facies. These are found as both large and small lenticular or block-like masses along fracture zones (mainly overthrust fault zones) within the Franciscan Formation. The highest grade of metamorphism seen in the Franciscan Formation itself is glaucophane schist facies (Cowan, 1974*) (Figures 2.VI.3, 2.VI.4, 2.VI.5). The margins or in some cases the whole of these high grade metamorphic rock masses have frequently been altered to mineral assemblages of the glaucophane schist facies as in the Franciscan Formation around them. High-grade metamorphic rocks are not the only exotic masses found along the fracture zones in the Franciscan Formation. Blocks or lenticular masses of shale, chert, greywacke sandstone and altered basalt are intermingled with them. Serpentinized peridotite of an ophiolite complex is also commonly encountered, K–Ar dating indicates that the metamorphism which promoted recrystallization of the above high-grade rocks occurred during the late Jurassic. This is the oldest geological age indicated by fossils present and is older than the low-grade metamorphism of the Franciscan Formation.

The general term for the fracture zone in which lenticular and block-shaped masses are included is the French word '*mélange*' meaning 'a mixture'. This term was first used by E. Greenly in his study of Anglesey, North Wales (Greenly, 1919*), where the presence of glaucophane schist facies rocks was well known. He called the geologic unit, chiefly composed of fractured shales and variously sized lenticular fragments an 'Autoclastic *mélange*'. '*Mélange*' is a most convenient term to express a situation in which complex rocks of various types and sizes are found in fractured rocks, distributed over a fairly wide area. Hsü (1968*) used the word to describe rocks, with these characteristics, found in the Franciscan Formation. The *mélange* defined here, is distributed mainly along the eastern part of the Franciscan zone.

The history of *mélange* formation in the Franciscan Formation is believed to be as follows:

(1) With the subduction of the oceanic plate under the *American plate* and rapid deposition of the Franciscan Formation near the ocean trench, the Formation itself was subjected to high pressure metamorphism. Its lower sections metamorphosed up to eclogite facies.

(2) The Franciscan Formation which had been deeply subducted along the so-called Benioff zone was forced up by thrust movements along the western margin of the American plate, and also sheared, brecciated and squeezed up into the weakly or even non-metamorphosed part of the Franciscan Formation, together with fragments of oceanic basalts and mantle material (Suppe and Armstrong, 1972*; Cowan, 1974*; Suppe 1972*).

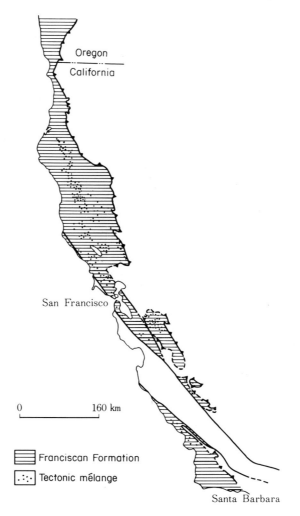

Figure 2.VI.3   Distribution of tectonic *mélange* (black dots) in the Franciscan
       Formation (after Coleman and Lanphere, 1971*)

After Hsü use of the term '*mélange*', it has been applied to only brecciated
and disturbed geologic structures, but Cowan suggested that the term
'tectonic *mélange*' should be applied to the products of the series of structural
movements, involving sedimentation, metamorphism, subsidence, thrusting,
fracturing, and 'squeezing up' that can be typically observed in the Franciscan
Formation.

A tectonic *mélange* thus defined is similar to an *olistostrome*, which is a
large-scale disturbed sedimentary product, but has quite a different meaning.

Figure 2.VI.4   Photograph showing an outcrop of tectonic *mélange* (after Cowan, 1974*)

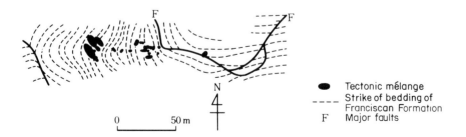

Figure 2.VI.5   Distribution of tectonic *mélange* (black) in the Franciscan Formation in the Ward Creek area, northern California (after Coleman and Lee, 1963*)

An olistostrome can sometimes become a tectonic *mélange*. The presence of an olistostrome has been firmly identified in the *mélange* of the Franciscan Formation. There are also examples in which, material squeezed up as the tectonic *mélange* has migrated again, as a result of a landslide on the ocean floor and has become incorporated into the sedimentary material being laid down to form an olistostrome (Figure 2.VI.6). Suppe (1972*) gave the name 'two way streets' model to a process whereby sedimentation proceeds along a long narrow belt centring on an ocean trench, so that lower sedimentary material is deeply buried and suffers high-grade metamorphism and then is squeezed up to form a sandwich, lens or block-shaped body along a thrust type fault developed at higher levels in the sediments (Figures 2.VI.6 and 2.VI.7).

The mechanism whereby the tectonic *melange* was created, in the Franciscan Formation, is fundamentally dependent on the relative motion between the plate forming the American continent and an oceanic plate that was being subducted below it. Fossil evidence and K–Ar dating suggest that, excluding fossils contained in redeposited olistostrome material, both the Franciscan Formation and tectonic *mélange* in it generally become younger from the east to the west of the Franciscan zone. This would support the 'two way streets' model of Suppe.

Numerical studies and experiments have supported this model as a likely explanation for the geological processes occurring in the Franciscan zone (Figure 2.VI.8) (Turcotte and Oxburgh, 1972*). The depth of subduction shown by the isotherms in Figure 2.VI.8 is considerably less than that

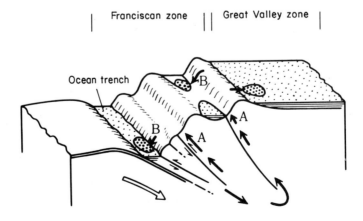

Figure 2.VI.6   Model demonstrating the formation of tectonic *mélange* and olistostrome in Franciscan Formation. A: squeezing up of the tectonic *mélange*. B: deposition of the olistostrome (after Cowan and Page, 1975*)

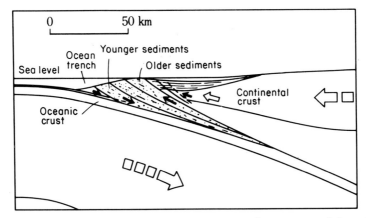

Figure 2.VI.7 'Two way streets' model, i.e. younger sediments are subducted from the ocean trench, under the continental plate, where they are compressed and subjected to high pressure metamorphism. They are subsequently squeezed up again along a plane of contact with the overthrusting continental plate. Same horizontal and vertical scales (Ernst, 1975*)

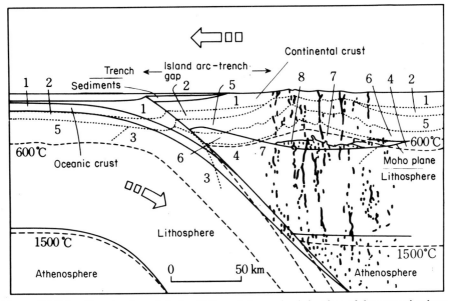

Figure 2.VI.8 Structural model of colliding plates and subduction of the oceanic plate underneath the continental plate (Ernst, 1973*) together with the vertical distribution of metamorphic facies (Ernst, 1974*). Isotherms after Turcotte and Oxburgh (1972*). Same horizontal and vertical scales

1. Zeolite facies
2. Prehnite-pumpellyite facies
3. Glaucophane schist facies
4. Eclogite facies

5. Greenschist facies
6. Epidote-amphibolite facies
7. Amphibolite facies
8. Granulite facies

calculated on the assumption that a 100 km thick oceanic plate was sinking at a rate of about 5 cm/year. However, this figure indicates that the Franciscan Formation must have undergone regional metamorphism at 150–300 °C and 5–8 kb, conditions which can be estimated from petrographic analysis of the metamorphic mineral assemblages, using experimental and theoretical phase equilibrium data.

(c)   The Klamath-Sierra zone and Great Valley Formation

The zone to the east of the Franciscan zone is composed of Precambrian to Jurassic, sedimentary material, their metamorphic equivalents, schists and gneisses, and huge quantities of granite. This is the so-called *Klamath-Sierra complex*, it forms a zone 400 km wide and 4000 km long, and extends from Mexico, through California to North Oregon. There may also be thin, late Jurassic and Cretaceous, strata in the zone and also Mesozoic, rhyolitic or andesitic, marine volcanics.

Isotopic and field studies have shown that the granitic rocks in the zone can be divided into three groups. The oldest, formed at 210–180 Ma (i.e. from late Triassic to early Jurassic), are distributed in the eastern half of the Klamath-Sierra zone. The second group, formed between 130–150 Ma (Late Jurassic) are found in the western half of the zone, whereas the youngest rocks, intruded between 60–90 Ma (Late Cretaceous) form a narrow belt about 40 km wide in the centre. The late Jurassic granitic intrusions are closely related to the Nevadan orogeny, which produced high temperature/ low pressure regional metamorphism and intensely folded the Jurassic strata and earlier sedimentary rocks.

At the same time (60–150 Ma), the deposition of extremely thick sediments with a rapidly subsiding geosynclinal basement, basaltic submarine volcanic activity, serpentinite intrusions and low temperature/high pressure meta-morphism, were all taking place within the Franciscan geosyncline, which developed parallel and to the west of the Klamath-Sierra zone (Figure 2.VI.8). The area trending NNW–SSE between the Franciscan zone and the Klamath-Sierra zone is occupied by the Great Valley Formation. This was deposited at about the same time as the Franciscan Formation and extended into the Cenozoic. The maximum thickness (18 000 m) of the Great Valley Formation principally composed of sandstone and shale is considerably less than that of the Franciscan Formation. Marine basaltic volcanic extrusives, cherts or limestones are not found suggesting that the Formation accumulated on the continental shelf or in the so-called arc-trench gap zone. The formation increases in thickness in a westerly direction, showing that the continental slope, which formed part of the basement of the Great Valley Formation, sank rapidly during the Jurassic and Cretaceous along with rapid subsidence of the Franciscan geosynclinal trough. It is also noteworthy that part of the western half of the Great Valley Formation was laid down directly on rocks with features characteristic of the Jurassic oceanic crust (Suppe and Armstrong, 1972*). The geological structure of the Formation is rather simple, with large scale open folds and very little faulting. The formation as a whole has suffered only low-grade diagenetic metamorphism.

Between the Great Valley formation and the Franciscan zone there is a large scale fault, which extends from southern Oregon to Southern California. This fault, known as the Coast Range Overthrust, is, in part, very gently inclined to the east and is a reverse fault along which the Great Valley Formation, on the east is thrust over the Franciscan Formation. At least three periods of displacement have been recorded since the Mesozoic, and the fault is believed to have been originally a 'Benioff' zone (Ernst, 1970*) of Mesozoic age.

Deposition of the Great Valley Formation began in the late Jurassic and continued for some 130 Ma, until the late Tertiary. The rate of deposition (rate of subsidence of the geosynclinal basement) was less than one-third that of the Franciscan Formation. From the Cretaceous, a change of sedimentary environment, from oceanic to continental, started on the eastern side of the basin where the formation was accumulating, and about 80 per cent of the late Tertiary sediments in the uppermost part of the Great Valley Formation are continental in origin. The small amount of volcanic material found in the formation is tuffaceous and andesitic to dacitic in nature, erupted at almost the same time as the intrusion of Cretacous granitic magma in the Klamath-Sierra zone. The Great Valley Formation contains no peridotite or serpentinite.

The Mesozoic history of the Californian mobile belt produced two belts with quite different sedimentary, metamorphic and igeous characteristics, in a large-scale, sheared, subduction zone formed between an oceanic plate moving from the west and a continental plate.

A common phenomenon encountered during the late Palaeozoic to late Mesozoic orogenic history of the circum-Pacific region, is the formation of thick sedimentary deposits, rich in basaltic volcanic material and with low temperature/high pressure metamorphism on the oceanic side. At about the same time on the continental margin, relatively thin sedimentary deposits, poor in volcanic material were formed and granitic igneous activity and high temperature/low pressure metamorphism were taking place.

The marked contrasts between the Ryoke belt and Sanbagawa belt in Japan, and between the Alpine zone and the Hokonui zone of New Zealand are very like the contrast between the Franciscan zone and Great Valley–Klamath–Sierra Nevada zone in California. Because there are some considerable differences between California and Japan or New Zealand in terms of the age of the sediments, the progress of metamorphism, the degree of deformation and other features (Ernst *et al.*, 1970*), it is difficult to apply the model of the

Californian tectonic development, just as they are, to other areas. However, it can safely be said that the geological, geochemical and geophysical research in the Californian mobile belt, including detailed study of the geological structure, fossil evidence, dating by isotopic methods, petrological investigation of metamorphic rocks, geophysical work and numerical analyses of the process of continental and oceanic plate collision and oceanic plate subduction have given us the most reliable and vivid model of the geodynamic development of 'Pacific type orogeny' as defined by Matsuda and Uyeda (1971*).

Yotaro Seki

References

Atwater, T. (1970): Implications of plate tectonics for the Cenozoic tectonic evolution of western North America, *Geol. Soc. Am. Bull.*, **81**, 3513–36.

Bloxam, T. W. (1956): Jadeite-bearing metagraywackes in California, *Am. Mineral.*, **41**, 488–96.

Coleman, R. G. and Lanphere, M. A. (1971): Distribution and age of high-grade blue-schists, associated eclogites, and amphibolites from Oregon and California, *Geol. Soc. Am. Bull.*, **82**, 2397–412.

Coleman, R. G. and Lee, D. E. (1963): Glaucophane-bearing metamorphic rock types of the Cazadero area, California, *J. Petrol.*, **4**, 260–301.

Cowan, D. S. (1974): Deformation and metamorphism of the Franciscan subduction zone complex, northwest of Pacheco Pass, California, *Geol. Soc. Am. Bull.*, **85**, 1623–34.

Cowan, D. S. and Page, B. M. (1975): Recycled Franciscan material in Franciscan mélange, west of Paso Robles, California, *Geol. Soc. Am. Bull.*, **86**, 1089–95.

Ernst, W. G. (1970): Tectonic contact between the Franciscan Melange and the Great Valley sequence—Crustal expression of a Late Mesozoic Benioff zone, *J. Geophys. Res.*, **75**, 886–901.

Ernst, W. G. (1973): Blueschist metamorphism and P–T regimes in active subduction zones, *Tectonophys.*, **17**, 255–72.

Ernst, W. G. (1974): Metamorphism and ancient continental margins, *The Geology of Continental Margins*, C.A. Burk and C. L. Drake (eds), Springer, Berlin, 907–19.

Ernst, W. G. (1975): Systematics of large-scale tectonics and age progressions in Alpine and circum-Pacific blueschist belts, *Tectonophys.*, **26**, 229–46.

Ernst, W. G., Seki, Y., Onuki, H. and Gilbert, M. C. (1970): Comparative study of low-grade metamorphism in the California Coast Ranges and the Outer Metamorphic Belt of Japan, *Geol. Soc. Am. Mem.*, *No. 124*, 1–276.

Greenly, E. (1919): *Geology of Anglesey, Mem. Geol. Surv. G. B.*, 2 vols.

Hsü, K. J. (1968): Principles of mélanges and their bearing on the Franciscan Knoxville paradox, *Geol. Soc. Am. Bull.*, **79**, 1063–74.

Matsuda, T. and Uyeda, S. (1971): On the Pacific type orogeny and its model—extension of the paired belts concept and possible origin of marginal seas, *Tectonophys.*, **11**, 5–27.

Suppe, J. (1970): Offset of late Mesozoic basement terrains by the San Andreas fault system, *Geol. Soc. Am. Bull.*, **81**, 3253–58.

Suppe, J. (1972): Interrelationships of high-pressure metamorphism, deformation and
  sedimentation in Franciscan tectonics, U.S.A., 24th IGC, Section 3, 552–9.
Suppe, J. and Armstrong, R. L. (1972): Potassium-Argon dating of Franciscan
  metamorphic rocks, *Am. J. Sci.*, **272**, 217–33.
Turcotte, D. L. and Oxburgh, E. R. (1972): Mantle convection and the new global
  tectonics, *Ann. Rev. Fluid Mech.*, **4**, 33–68.

VII *The Canadian Shield and aulacogens*

Precambrian granitic rocks are widely distributed in the centre of continents,
such as in the *Canadian Shield*, on the North American continent and the
Russian or the Siberian platforms, on the Eurasian continent. Structures
known as 'aulacogens', found in such areas, are graben-like fracture zones
several tens of kilometres wide and several hundreds of kilometres long.
When these structures developed, long narrow depressions bounded by
nearly vertical faults on both sides were formed, initially accompanying
volcanic activity, and subsequently filled in by huge quantities of sedimentary
material.

The word 'aulacogen' is derived from the Greek words meaning 'channel/
ditch' and 'genesis'. It was used for the first time by the Russian geologist
Schatsky (1964). He had noticed the presence of huge graben-like structures
on the Russian platform, buried by Palaeozoic strata, Schatsky described a
geological structure in *Dnepr-Donetsk* to the south of Moscow and proposed
that it should be called an 'aulacogen' to distinguish it from other structures,
because the history of the area had features quite distinct from those of
mobile belts or sedimentary basins in cratonic areas. So, what are the special
characteristics of aulacogens? What is their relationship to orogenic zones and
other geological structures? With what type of areas visible on earth today
can their history be compared? Schatsky died before he could summarize all
his ideas about these structures, but his work was continued by his former
students and colleagues, Bogdanov and others. Aulacogens are described as
follows: 'They are huge linear graben running for several hundreds of
kilometres across the continental crust in platform regions and buried by thick
sedimentary deposits. Volcanic activity occurs in the early stages of aulacogen
development, and their sedimentation phase resembles that of a miogeosync-
line. Weak metamorphism may also be found in places.' (Kosygin, 1969*.)
Such structures were subsequently identified all over Russia, and the term
'aulacogen' came into common use.

The morphology of aulacogens and one explanation of their tectonic
history have been discussed by Schmidt-Thomé (1972*). He took aulacogens

to be geological structures showing the features of an early phase in geosynclinal development, or as structures seen in transition zones between geosynclines and cratons. In contrast to this, Salop and Scheinmann (1969*) indicated that aulacogens are graben zones formed in a cratonic region, and joined at their ends to a geosyncline surrounding the craton. The North American geologists Hoffman (1973*), Hoffman *et al.* (1974*) and Burke and Dewey (1973*) provided the further explanation that, because aulacogens are associated with orogenic zones in either T or Y configurations, the geosynclines and aulacogens had developed from former *triple junctions*. Their explanation runs as follows: a mantle plume rising from deep in the earth pushed up the continental crust and fractures it, forming a tri-directional fracture. In two of these directions, fracture development is considerable, forming an ocean and dividing the continent into two; the ocean is subsequently closed again, and an orogenic zone develops; in the remaining direction, however, the fracture extends through the craton interior, retaining its graben-like form, and is gradually filled with sedimentary material. This remaining 'failed arm' is then the aulacogen. They cited the *Athapuscow aulacogen* and Coronation geosyncline from the Canadian shield of Proterozoic age as good examples of the relationship between aulacogens and orogenic zones. Hoffman (1969*, 1973*, 1974*) and Hoffman *et al.* (1974*).

(a)   The geology of the northwest Canadian Shield

The Precambrian rocks, which are extensively exposed throughout northwest Canada, can be divided into the Slave, Churchill and Bear Provinces as shown in Figure 2.VII.1 according to their structure and radiometric age.

The ages of the volcanic and plutonic rocks, found in the Slave Province, are largely between 2600–2300 Ma, and the principal diastrophism occurred at 2480 Ma. The oldest age given by Rb-Sr radiometric dating, for rocks in this area, is 3000 Ma, obtained from granitic gneiss in the northwest of the Slave Province (Frith *et al.*, 1977*).

In the Churchill Province, ages of 2600–1600 Ma have been recorded, and there is no doubt that there are also Archaean rocks in part of this area. By the early Proterozoic period, both these areas had completely consolidated and developed into cratons, called, by Hoffman (1973*) the Slave craton and the Churchill craton, respectively.

Early Proterozoic rocks are found in the Bear Province, their ages being mainly between 2100–1600 Ma. Similar, early Proterozoic rocks are also found around the eastward extending arm of the Great Slave Lake. This is the area named the 'Athapuscow aulacogen' by Hoffman (1973*). The topography and strata distribution around these areas are dominated by the McDonald fault which runs ENE–WSW. This fault, with its excellent continuity, is still clearly recognizable today. The age (1700–1800 Ma) of muscovites, found

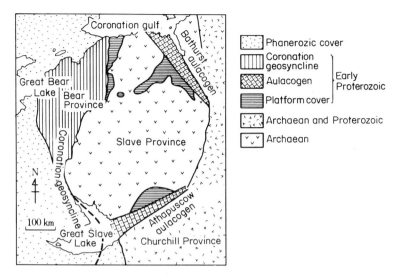

Figure 2.VII.1   Geological map of the northwestern Canadian Shield (Hoffman, 1973*)

in sheared granite along the fault, and the structures in its western extension, suggest that movement on the fault took place during the mid-Proterozoic and the Palaeozoic. However, the principal displacement occurred in the early Proterozoic when the aulacogen developed.

South from Coronation Bay, Bathurst Inlet projects into the Slave Province. The topography of this area and features of the rocks found there, have given rise to the view that this is an area similar to the Athapuscow aulacogen (Hoffman 1973*, Burke and Dewey 1973*).

Numerous diabasic dykes run parallel to each other in a NNW–SSE direction over a huge area covering some 1000 km E–W, and 1600 km N–S from the MacKenzie region of northern Canada to Saskatchewan. The length of an individual dyke may sometimes reach 250 km. Each dyke cuts virtually all the Precambrian rock formations in the Bear, Slave and Churchill Provinces almost vertically and moreover, in straight lines. They are estimated to have intruded at around 1300 Ma ago. Precambrian movements in NW Canada had all ceased before the intrusion of the MacKenzie dykes.

(b)   Coronation orogenic belt

Early Proterozoic rocks are distributed in an almost north–south direction in the Bear Province, judging from the lithology, thickness and sedimentary structures seen there, and one may conclude that this area was formerly in the same kind of environment as a so-called geosyncline. At the beginning of the Proterozoic, such a sedimentary basin, called the Coronation geosyncline, developed extensively around a continent composed of Archaean rocks.

The Proterozoic rocks and strata in the Bear Province are divided, from the west, into the Great Bear batholith and volcanic belt, the Hepburn batholith and metamorphic belt, a thrust zone and a platform zone (Figure 2.VII.2).

The Great Bear batholith and volcanic belt is composed, principally, of volcanic and plutonic rocks and is connected to the Hepburn batholith and metamorphic belt to the east by a steep angled fault. Mantled gneiss domes are found in the south of the Hepburn batholith, near the Slave Province, amongst the metamorphic rocks derived from conglomerate, sandstone, mudstones, and dolomite, etc. Rock formations in the Hepburn batholith and metamorphic belt are extremely folded and show a remarkable foliation. The clastic facies tend to thicken towards the west, and are estimated to be about 7000 m thick. The amounts of both dolomite and igneous rock in the sedimentary complex increase to the east, whilst the clastic strata become thinner. The major source of supply of clastic material in these formations, formed in the early and middle stage of deposition, lay to the east, but sedimentary structures in clastic rocks laid down subsequently suggest a palaeocurrent, west to east.

Many low-angle thrust faults are found in the thrust zone, in the east of the Hepburn batholith and metamorphic belt. These were all formed by thrust

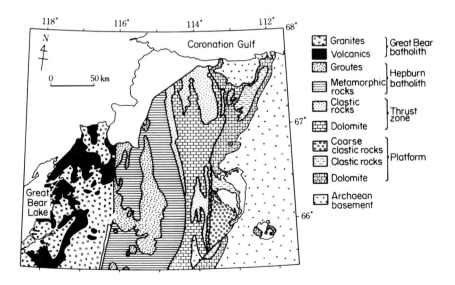

Figure 2.VII.2   Geological map of Bear region (Hoffman, 1973*)

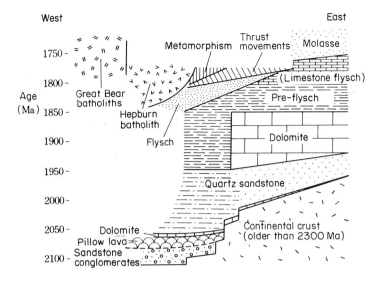

Figure 2.VII.3   History of Coronation orogenic zone (Hoffman, 1973*)

movements from west to east. However, the strata are hardly deformed in the platform zone further to the east. The majority are of continental nature covering, unconformably, the Archaean rocks. The history of the Coronation geosyncline has been summarized by Hoffman (1973*) as in Figure 2.VII.3. This Proterozoic geosyncline has much in common with the orogenic belts which developed along the western margins of the North American continent during the Palaeozoic and Mesozoic. These similarities include;

(a) the total structure, i.e. rock distribution, foliation and direction of fold axes, all run almost north–south parallel to the continental margin;

(b) from west to east, the tectonic belts are in the order: metamorphic zone, folded zone, overthrust zone and cratonic platform;

(c) the sedimentary strata are thicker to the west and thinner to the east;

(d) an upheaval of a granitic zone supplied clastic material at the end of the Mesozoic, and was followed by the formation of low angle thrust faulting from west to east.

(c)   Athapuscow aulacogen

The Great Slave Super group is the name applied to the rock formations of the island-dotted, eastern arm of the Great Slave Lake. An Rb-Sr age of 1872 ±10 Ma has been obtained for spilite basalts within the Supergroup (Baadsgaard *et al.*, 1973*), and suggests that the Supergroup was formed between 2200–1800 Ma.

The Great Slave Supergroup can broadly be divided into upper and lower sections. The lower section is 2000–5000 m thick, with a predominance of shallow marine sandstones and limestones, but it becomes thinner towards the northeast, and is of cratonic type. In contrast, in the southwest deep marine shales and turbidites are found and are cut by volcanic rocks, such as the spilite basalts mentioned above. The direction from which clastic material was supplied, at this time, was from NNE to SSW, from the interior of the continent towards the margins. In the upper section of the Great Slave Supergroup the direction of supply is completely reversed, being SSW to NNE, or towards the interior of the continent. It is 2000–3000 m thick and composed mainly of alternating layers of turbidites, limestones and mudstones, changing into shallow marine types higher up the succession. Incorporated in the clastic rocks are fragments of older sandstones, shale, limestones and volcanics, commonly seen in orogenic zones, and the area from which such clastic material was supplied may thus be deduced to have been an orogenic zone.

Coarse clastic formations, in places up to 5000 m thick, developed in the area of the Athapuscow aulacogen, covering the Great Slave Supergroup in an angular unconformability. They are largely of continental origin and contain fanglomerates and landslide material, but in part are interspersed with lacustrine sediments. The basins, where this material was deposited, are bounded by steep angled faults, and they were obviously formed in a period of extensive block movements. Locally, they also contain basic volcanic lava, leaving no doubt that volcanic activity accompanied the block movements.

Let us take a look at a vertical section, cutting the Athapuscow aulacogen at right angles to its extension direction (Figure 2.VII.4(d)). The McDonald fault marks the southern boundary of the aulacogen and runs between the Archaean basement and the Proterozoic strata. A graben zone is found in the middle of the aulacogen, and thick layers of sedimentary and volcanic rocks, filling it, became folded, as a result of subsequent movement. However, some of the material deposited in the northern part of the graben zone, changed to a cratonic facies and spread over the basement.

Hoffman *et al.* (1973[*] and 1974[*]) presented a tectonic model to explain the history of the development of the Athapuscow aulacogen as shown in Figure 2.VII.4. Structural movements began with the uplift of the whole area, followed by the formation of a graben, then by gentle but regional sinking and finally severe compression and further fault block movements. The graben zone, formed by the early fault movements, extended almost due east–west, but opened slightly to the west and was perhaps linked to the Coronation geosyncline. As Dickinson (1976[*]) explains, a wedge-shaped aulacogen on the map, possesses different structures in its deeper parts. Its structure in the craton interior is reminiscent of a platform, whereas that of the continental margin has features characteristic of mobile belts. The miogeosynclinal

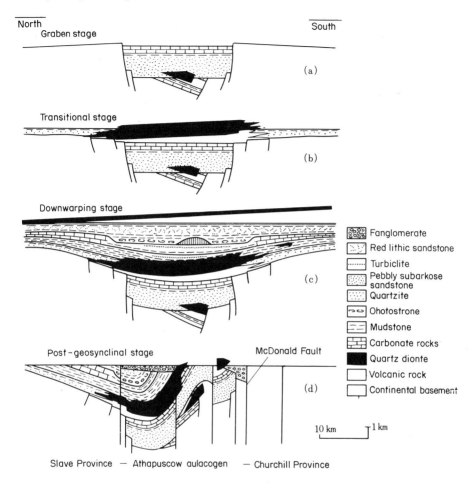

North   South
Graben stage

                                                                           (a)

Transitional stage

                                                                           (b)

Downwarping stage

Fanglomerate
Red lithic sandstone
Turbiclite
Pebbly subarkose sandstone
Quartzite
Ohotostrone
Mudstone
Carbonate rocks
Quartz dionte
Volcanic rock
Continental basement

                                                                           (c)

Post-geosynclinal stage   McDonald Fault

                                                                           (d)

10 km   1 km

Slave Province — Athapuscow aulacogen — Churchill Province

Figure 2.VII.4   Schematic cross-section showing evolution of Athapuscow aulacogen
(after Hoffman, Dewey and Burke, 1974*)

characteristics in the lower section of the Great Slave Supergroup, pointed out by Hoffman, could well reflect the environment around the junction between the Athapuscow aulacogen and the Coronation geosyncline.

(d)   Formation of aulacogens

The Coronation geosyncline and Athapuscow aulacogen were contemporaneous, forming over a long period. The two are united at the Great Slave Lake and their spatial relationship forms the letter Y as shown in Figure

2.VII.1. The Coronation geosyncline is a structure developing along the continental margin, and the boundary between the Bear and Slave regions may reasonably be regarded as the boundary between the craton and the sea during the early period of geosyncline formation. On the other hand, the Athapuscow aulacogen developed across the craton. Uplift subsidence and associated block movements were predominant here. The Slave and Churchill areas, separated by the aulacogen, perhaps showed only very slight relative movement. In other words, the Y-shaped junction seen today may show the form it possessed early in the period when both the Coronation geosyncline and the Athapuscow aulacogen were being formed. In the early Proterozoic, when the development of this region began, a huge Y-shaped fracture zone, centred on the Great Slave Lake was formed which subsequently developed into a aulacogen and geosyncline. Hoffman *et al.* (1974*) summarized and compared the history of the two regions as in Table 2.VII.1.

The Athapuscow aulacogen was the first place in North America to be studied and designated an aulacogen. However, this is not to say that there are no other examples of well-researched areas in which similar structural characteristics may be found. For instance, a further case is that of the Southern Oklahoma aulacogen.

Table 2.VII.1   Correlation of tectonic history of Coronation geosyncline and Athapuscow aulacogen

	Coronation geosyncline	Athapuscow aulacogen
Stage V	Postgeosynclinal stage (coarse sediments)	Postgeosynclinal stage (coarse sediments)
Stage IV	Compressional stage	Compressional stage
Stage III	Exogeosynclinal stage† (morasse, limestone flysch, flysch)	Downwarping stage (morasse, limestone flysch, flysch)
Stage II	Transitional stage	Transitional stage
Stage I	Miogeosynclinal stage (dolomite, quartz sandstone)	Graben stage (dolomite, quartz sandstone)

† Sediments were laid down and accumulated in areas such as the continental interior, margins and the ocean floor. The individual rock associations and thicknesses are characteristic of the various environments. Those formed at the continental margin are thicker than cratonic sediments, their lithofacies being more diverse and changeable. Subsequently, these areas often turn into orogenic zones, whilst a broad sedimentary zone may form between this orogenic zone and the craton and over the continental crust. The sedimentary material consists mainly of thick deltic sequences and turbidites, mostly supplied from the continental margins in response to the uplift of the core of the orogenic zone. Such sedimentary zones are sometimes called exogeosynclines. An example is the sedimentary zone formed to the west of the Appalachian Mountains in the early Palaeozoic.

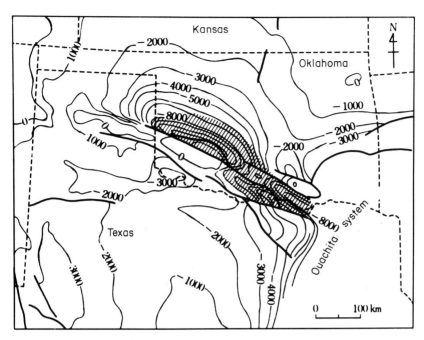

Figure 2.VII.5 Basement structure around Oklahoma. Basement rocks of 1000–1400 Ma mostly covered by Palaeozoic strata. Diagram shows height above sea level (m) of the top of the basement. Thick lines — faults. Shaded area — southern Oklahoma aulacogen (Ham, 1969*)

A Palaeozoic sedimentary basin in southern Oklahoma, known as intracratonic geosyncline, used to be called the 'southern Oklahoma geosyncline' (Ham, 1969*). This area is delineated by steep faults as shown in Figure 2.VII.5 and is long, but narrow and deep. It was filled with sediments, up to a thickness of 12 000 m, during the late Cambrian to the Permian, deposited over mid-Cambrian acidic volcanic rocks more than 2000 m thick. In this period, there was also an oceanic region extending from Arkansas to southern Oklahoma and western Texas known as the Ouachita geosyncline. This was where the *Ouachita system* was formed, characterized by Devonian chert and carboniferous and Permian turbidites. The southern Oklahoma geosyncline was born in a continental crust composed of Proterozoic rocks. It was wedge-shaped, opening out slightly towards the southeast, where the Ouachita geosyncline is extensively developed. Severe block faulting also occurred in the southern Oklahoma geosyncline—Burke and Dewey (1973*) and Hoffman *et al.* (1974*) who studied these areas recognized that the southern Oklahoma geosyncline possessed structural features similar to those of the aulacogens described above and named it the '*southern Oklahoma aulacogen*'.

The relationship between the Coronation geosyncline and the Athapuscow aulacogen seen in the Proterozoic of northwest Canada, closely resembles that of the Ouachita geosyncline to the southern Oklahoma aulacogen in the Palaeozoic of southern USA. With respect to the arrangement of orogenic zones and their associated aulacogens their structure resembles the RRR type of triple junction with three-armed radial rifts of the various types of *triple junction* discussed by McKenzie and Morgan (1969). Burke and Dewey (1973*) and Hoffman *et al.* (1974*) explained the relationship between aulacogens and orogenic zones with the processes shown in Figure 2.VII.6. Localized crustal doming occurs above a mantle plume, and the formation of three radial rifts and the subsequent expansion of two of them leads to the development of an ocean basin. This basin finally becomes a folded zone in the continental margin, whilst the remaining abandoned rift is filled with sediments and further loaded with clastic materials derived from an advancing orogen.

Burke and Dewey (1973*) pointed out that geological structures created in this way could be found all over the world, dating from various periods. They cited 30 examples from the Red Sea – Gulf of Aden – Afar Depression system, which are still very clearly marked today, to the Proterozoic Athapuscow aulacogen (see Figure 2.1 on page 14). The most typical example of modern analogues is perhaps the relationship between the Benue Delta in Africa and the Atlantic Ocean. This delta developed in Nigeria with the infilling of one of three radial rifts, formed between 120–80 Ma ago; the other two rifts having expanded to become part of the present Atlantic Ocean.

The configuration of aulacogens and orogenic zones suggests the presence of 'fossil' triple junctions, and provides us with an important clue in the study of *plate tectonic models* in the past. By analysing the whole history pertaining to such structures we can probably reconstruct world-wide structures that existed in ancient periods. The existence of a huge continental crust and mantle plume rising from beneath to fracture it, are essential to the formation of aulacogens and orogenic zones. Hoffman (1974*) and Burke (1977*) said that no structure resembling the Athapuscow aulacogen and associated Coronation orogenic zone discussed here, are found within Archaean rocks. If this is true, then the type of movements which led to the creation of triple junctions; expansion of the oceanic basins and continental drift, followed by the formation of orogenic zones, does not seem to have begun until the Proterozoic some 2500 million years ago.   Vigorous igneous activity invariably takes place in the early to middle stages of the development of aulacogens. This is rather a continental characteristic and differs markedly from so-called ophiolites. Mineralization is frequently associated with this

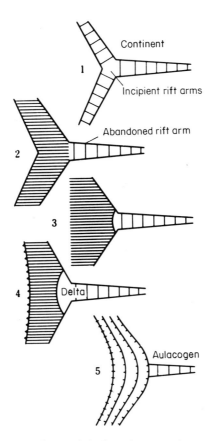

Figure 2.VII.6   Development from triple junction to aulacogen and orogenic zone (Dewey and Burke, 1974*). 1. Three-armed radial rift system. 2. Two arms spread to produce a narrow ocean basin. 3. Sedimentation at the continental margin and infilling of abandoned rift. 4. Closing ocean with subduction along trench. 5. Development of folded zone in continental margin and final configuration of aulacogen

activity leading to the formation of metal ore deposits such as copper or zinc. In the later stages of aulacogen development huge deltas form in the area where the aulacogen and the oceanic basin join. The sediments found there are relatively fine-grained and contain organic matter. Since the rate of deposition is high and its period long, organic matter accumulates in the sediments and may later turn into fossil fuels. Aulacogens are thus of great interest, not only as specialized geological structures, but also as areas generating considerable natural resources. Schatsky, in fact, began his research into aulacogens at the beginning of the Second World War, and is said to have been ordered to do so as part of the quest for oil resources in the Soviet Union.

Shinjiro Mizutani

References

Baadsgaard, H., Morton, R. D. and Olade, M. A. D. (1973): Rb-Sr isotopic age for the Precambrian lavas of the Seton formation, East Arm of Great Slave Lake, Northwest Territories, *Can. J. Earth. Sci.*, **10**, 1579–82.

Burke, K. (1977): Aulacogens and continental breakup, *Ann. Rev. Earth Planet. Sci. 1977*, **5**, 371–96.

Burke, K. and Dewey, J. F. (1973): Plume-generated triple junctions: key indicators in applying plate tectonics to old rocks, *J. Geol.*, **8**, 406–33.

Dickinson, W. R. (1976): Plate tectonic evolution of sedimentary basins. In Plate Tectonics and Hydrocarbon Accumulation, *Short Course*, AAPG Contin., Education Comm. and AAPG Res. Comm. 1–56.

Frith, R., Frith, R. A. and Doig, R. (1977): The geochronology of the granitic rocks along the Bear-Slave structural province boundary, northwest Canadian shield, *Can. J. Earth Sci.*, **14**, 1356–73.

Ham. W. E. (1969): *Regional Geology of the Arbuckle Mountains, Oklahoma*, Oklahoma Geol. Survey, Guide Book XVII, 52 pp.

Hoffman, P. F. (1969): Proterozoic paleocurrents and depositional history of the East Arm fold belt, Great Slave Lake, Northwest territories, *Can. J. Earth Sci.*, **6**, 441–62.

Hoffman, P. F. (1973): Evolution of an early Proterozoic continental margin: the Coronation geosyncline and associated aulacogens of the northwestern Canadian shield, *Phil. Trans. R. Soc. London*, A273, 547–81.

Hoffman, P. F. (1974): Shallow and deep water stromatolites in lower Proterozoic platform-to-basin facies changes, Great Slave Lake, Canada, *Am. Assoc. Petrol. Geol. Bull.*, **58**, 856–67.

Hoffman, P. F., Dewey, J. F. and Burke, K. (1974): Aulacogens and their genetic relation to geosynclines, with a Proterozoic example from Great Slave Lake, Canada, *SEPM Sp. Publ.*, No. 19, 38–55.

Kosygin, Yu. A. (1969): *Tektonika*, 616 pp. Nedra, Moscow.

Salop, L. I. and Scheinmann, Y. M. (1969): Tectonic history and structures of platforms and shields. *Tectonophys.* 7, 565–97.

Schmidt-Thomé, R. (1972): *Lehrebuch der Allgemeine Geologie, Bd. II. Tektonik*, 579 pp., Ferdinand Enke, Stuttgart.

VIII　*Salt domes in the Gulf Coast*

The Mississippi River flows for more than 6000 km, from its source in Missouri, to the Gulf of Mexico, forming an enormous delta at its mouth, surrounded by a broad plain. Five small hills, a few kilometres across and up to 50 m high, are aligned along the Louisiana coast in a line 75 km long, striking almost N 40° W. These are collectively called the Five Islands of Louisiana, and comprise from the northwest Jefferson Island, Avery Island, Weeks Island, Cote Blanche Island and Belle Isle. The highest point in southern Louisiana is known as the Devil's Backbone, but despite this name, it is a mere 52.1 m above sea level. It forms the spine of the largest of the hills, Weeks Island.

It was on Avery Island in 1862 that the chance discovery was made of rock salt directly below the strange little hills in this flat lowland. The first oil in southern Louisiana was discovered in 1901 from the Jennings Hill of similar topography about 65 km northwest of the five islands. As the close relationship between such topographical features, rock salt, and then oil became clearer, research into the geological structures of these oil producing areas accelerated rapidly. As a result huge dome-shaped salt masses were found underground, and it was realized that they had risen up to the surface from deep below and had pushed up the surrounding strata. Round hills such as the five islands are a topographical reflection of salt domes, which are still rising.

Geophysical investigations such as gravity surveys or seismic refraction surveys have indicated salt domes at depth, in places where there is no topographical manifestation of their presence. Their presence has been confirmed by deep drilling, as the natural gas and oil reserves have successfully been exploited. Numerous salt domes have been found in southern Louisiana and in the other Gulf Coast area, enclosed as shown in Figure 2.VIII.1 (Spillers, 1969*). More than 300 salt domes are known, between

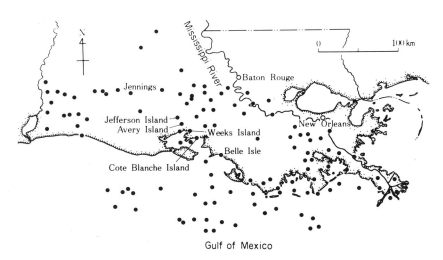

Figure 2.VIII.1   Salt domes in southern Louisiana (Spillers, 1969*)

Alabama, Mississippi, Louisiana and Texas in the southern United States and Verracruz and Tabasco in Mexico.

Salt domes are not just valuable as a source of rock salt for the chemical industry, but are also of great interest in themselves as they form the geological structures which are associated with oil and natural gas deposits. The phenomenon whereby a certain rock mass seems to be squeezed up from below and push up the overlying strata has long been referred to as diapirism (Braunstein and O'Brien, 1968*). The Swedish chemist Arrhenius (1912*), who was interested in the *salt domes* of northern Europe, was the first to point out that the rise of salt into overlying formations, i.e. the formation of *diapirs*, is due to the *buoyancy* attributable to the salt's relatively low specific gravity. If less dense strata is overlain by relatively dense strata, the lighter material tend to rise upwards. If the underlying less dense material is capable of plastic flow it will begin to rise deforming the surrounding rock formations. Series of structural movements, with such origins, have been identified widely throughout northern Germany and are known as 'Salstektonik'. Trusheim (1957*) studied these movements, and used the name of '*Halokinese*', i.e. *salt tectonics*, to distinguish them from orogenic movements originating in lateral compression. Here we will discuss a series of structural problems associated with the development of salt domes, citing examples found in the southern part of the North American continent.

(a)   Weeks Island salt dome

A topographical map of Weeks Island, one of the Five Islands in Louisiana coast, is shown in Figure 2.VIII.2. It is a small hill about 3 km across and just over 50 m high. The underlying salt dome is also about 3 km across and the centre of the dome corresponds almost exactly with the centre of this round hill. A small valley runs north–south, slightly east of centre, and is said to mark the surface expression of a shear zone developing inside the salt dome.

The shape of the salt dome as a whole has been ascertained by the geophysical survey and directly by drilling. The subsurface contours of the dome in Figure 2.VIII.3 and the geological profile in Figure 2.VIII.4 illustrate its external shape to be virtually that of a round pillar. However, salt domes do not always go straight up; sometimes the top spreads out laterally and overhangs its lower part. The head of the Weeks Island dome reaches to a mere 26 m below the ground surface, and it is known to extend downwards at least 5 km. The salt domes of the Five Islands are all believed to originate from a mother salt bed, known as Louann Salt, located 16–19 km below ground, but it has not been established whether all the domes are continuous to that depth. If they are connected, their external form will be similar to that of long thin fingers, with diameter to height ratios of 1:5 to 1:6.

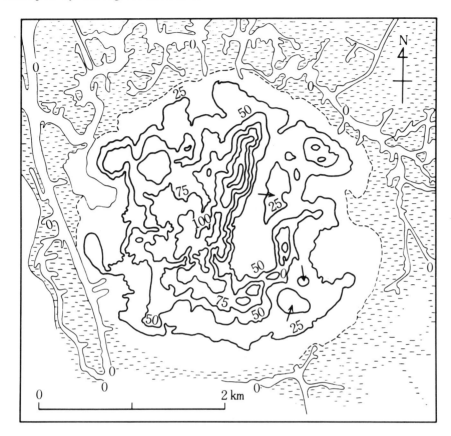

Figure 2.VIII.2 Topograhy of Weeks Island (after Morgan, 1976*). Contours at 25 ft intervals, stippled area — marshlands

Tunnels about 20 m wide, 20 m high and 1000 m long have been dug in this dome from E–W, N–S. The ceiling, walls, indeed the whole tunnels are composed of semi-transclucent rock salt crystals. The illuminated interior glitters and is as beautiful as a snow palace. A square pillar, 30 m wide at the base, has been left between the tunnels, to prevent them from collapsing. The salt, from which the salt dome is made, is over 99.5% NaCl in its purer parts. There are uniform block-like areas in the salt, and also banded layers ranging from a few centimetres to several tens of centimetres wide. These result from a bedded structure composed of white bands with a high NaCl content and greyish ones containing a few per cent anhydrite (CaSO). Blackish bands are said to be due to small amounts of impurities in the anhydrite. These structures represent the alternating layers produced when the rock salt was

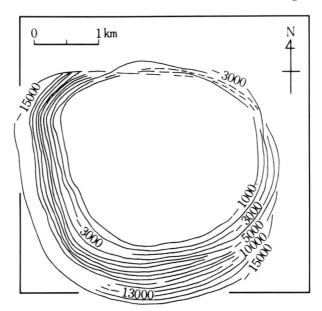

Figure 2.VIII.3   Subsurface contours illustrating the external shape of Weeks Island
salt dome (Spillers, 1969*). The contours crossing each other in the north indicate a
lateral overhang of the salt dome. Contours in feet. Boundary of diagram the same as
Figure 2.VIII.2

formed as evaporite. If one looks at and traces this bedded structure in a
specific direction, the transition from a white layer to a dark grey one forms a
clear plane; on the other hand, where a dark grey layer gives way to the next
white layer, the change is gradual. Kupfer (1962*) claimed that this type of
relationship shows a sequence of deposition similar to graded bedding and it
is helpful in assessing top and bottom relationships in stratiform salt beds.
This stratification may also be used as an aid in determining the internal
structure of the salt dome. Interestingly, a feature commonly observed in the
domes of the Gulf Coast area is that the stratification is inclined at an angle of
about 80°, but has a variable strike. This fact shows that the layers of salt are
folded inside the dome and that the fold axes are almost vertical. The graded
bedding enables one to identify whether a structure is synclinal or anticlinal
even if the fold axis is vertical. Particularly noteworthy in the configuration of
the folds is that closed curves are visible when the stratiform structure appears
on a horizontal plane. This is rather similar to the annual rings seen in the
cross-section of a tree trunk. If one tries to examine whether the structure is
concave upwards or downwards using graded bedding, it is found that these
closed curves are a horizontal cross-section of a spine-shaped closure made by

the stratiform structure. As illustrated in Figure 2.VIII.5, refolded structures in which a fold once formed is subjected to new deformation, and drag folding accompanying large scale folds are also commonly seen.

The geology of the surrounding area has been studied in detail with nearly 300 drillings, which have established the stratigraphy from the Holocene to the Oligocene. The perimeter of the salt dome is very steep, indeed almost vertical, with its top overhanging slightly towards the north. Deep down, at about 3000 m, a layer of shale is found, surrounding the salt dome like a collar. A shear zone exists between this shale and an overlying bed and the shale is also presumed to have flowed with the ascent of the salt. Strata surrounding the dome are more steeply inclined as they approach it laterally and their subsurface contours form concentric rings around the salt dome (Figure 2.VIII.6). A number of faults radiate out from the perimeter of the salt dome, but they are not linked to the shale band. The strata is more deformed at greater depths, as may be shown in the profile in Figure 2.VIII.4. Conversely, the shallower the position the more gentle the inclination of the strata. Natural gas and oil accumulate in the strata surrounding salt domes, especially in those highly permeable to water. Miocene sandstones form the principal oil bearing layers in the Five Islands region. The salt formation

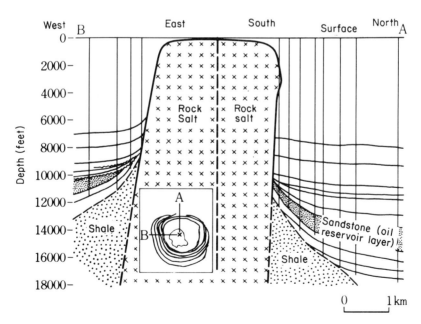

Figure 2.VIII.4   Profile of Weeks Island salt dome (Atwater and Forman, 1959*).
Vertical and horizontal scales equal. Fine vertical lines show bore holes

which abuts this steeply dipping layer is highly impermeable to the gas and oil. This, assisted by overlying strata produces the oil reservoir.

(b)   The process of salt dome development

Various hypotheses have been put forward concerning the origin of the hills dotted like islands in the southern Louisiana plain. These included; volcanoes, the remnants of natural dykes along the Mississippi River or the vestiges of islands created during the Cretaceous transgression. Some re-

0   100 m

Figure 2.VIII.5   Internal structure of Weeks Island salt dome (shaft roof 660 ft below sea level) (after Kupfer, 1962*)

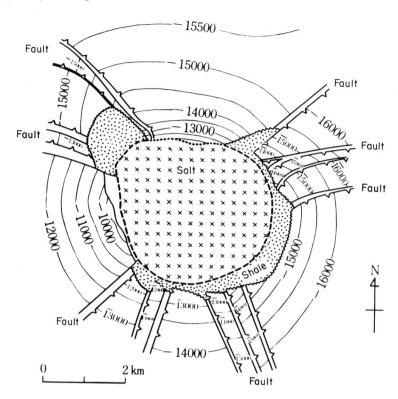

Figure 2.VIII.6   Subsurface structure of Weeks Island salt dome and its surrounding area (Atwater and Forman, 1959*). Subsurface contours refer to depth measured from upper edge of sandstone designated the 'S' layer. Units in feet

garded the fact that the Five Islands run in an almost straight line parallel to the direction of the Wichita-Arbuckle Mountain system in southern Oklahoma significant. They claimed that these topographical features were related to large scale orogenic movements. Such ideas mainly date from the nineteenth century when the presence of salt had not yet been demonstrated. After Nettleton (1934) conducted experiments on the phenomenon of buoyancy in fluid materials due to differences in their specific gravities the idea of Arrhenius (1912*) that the origins of salt domes lay in buoyancy became accepted in the United States and it was believed that the Texas or Louisiana salt domes were similar to those found in Germany or Romania.

The topography of the round Five Islands hills has thus been determined by the presence and gentle rise of enormous pillars of salt directly underneath them. However, there are many places where the presence of salt masses has no apparent effect on the landscape. Perhaps the salt is rising gradually from the depths of the crust even now and will ultimately create hills like Weeks Island on the surface. If upward growth of the salt formation continues, part of it can break through to the surface. In Iran, for example, salt is reported flowing out from the tops of mountains and forming salt glaciers (Kent, 1958*). That the salt masses are still moving can be observed in tunnels bored through them. We know of an example in Germany, where redrilling of a mineshaft is required every eighteen months and of a further example in Texas, where plastic flow in a tunnel reaches 0.9 mm per year. Because of the diverse forms of salt formations and the fact that some are still moving, we can compare the structures encountered throughout the world and understand a developmental sequence based on their morphological similarities and differences. This suggests a process whereby a salt dome grows from the mother bed and finally becomes a salt glacier (see Chapter 5.9, Figure 5.34).

As described earlier, spine-shaped closures and refolded structures can often be seen inside salt domes. These features show that the upwards movement of the salt does not proceed uniformly throughout the mass but takes place in several separate sections. The internal structure is indicative of a mechanism whereby the inside of the salt dome is separated into different spines and the whole structure moves upwards with repeated rises of these relative to each other. Such differential movement probably creates the shear zones inside the salt dome but these are rarely recognizable as faults because the shear zones themselves again flow and deform.

Surrounding sedimentary strata is dragged upwards and bent by the movement of the salt, and may sometimes be deformed to the extent that they become vertical or locally overturned. The deformations only develop around the perimeter of the salt dome. Both their scale and shape are closely related to the size of the dome, and in places without salt formations there is virtually no deformation of the strata. In brief, even in areas subjected to no regional structural movements whatsoever, the presence of thick salt layers and of plastic flow within them, can result in the local formation of independent small folded structures.

The minerals which make up salt domes have the densities shown in table 2.VIII.1. The specific gravity of halite (rock salt) is clearly low compared with those of rock-forming silicate minerals. The values measured for rock salt in salt domes is in the range $2.0 - 2.2$ g/cm^3; this is slightly lower than for pure NaCl because of the presence of small amounts of sylvite and pores between the minerals. The compressibility of halite is about $(V_o - V)/V_o = 4 \times 10^{-6}$/ bar ($V_o$ and V represent volume prior to and post compression respectively) and there is therefore little change in its density up to depths of about 10 000 m. In contrast, the density of sedimentary rock is about 1.6 g/cm^3 in its unconsolidated state. As it becomes buried and compressed, so its density probably rises as shown in Figure 2.VIII.7, we see that the density of sedimentary rocks will become greater than that of halite at depths in excess of 900 m.

Table 2.VIII.1   Densities of minerals in salt layers and principal crust materials

Carnalite	$KMgCl_3.6H_2O$	1.602 g/cm³
Sylvite	KCl	1.984 g/cm³
Kainite	$KMg(SO_4)$ Cl.3H₂O	2.131 g/cm³
Halite	NaCl	2.165 g/cm³
Gypsum	$CaSO_4.2H_2O$	2.317 g/cm³
Anhydrite	$CaSO_4$	2.960 g/cm³

Sandstone	2.17–2.70 g/cm³	Eclogite	3.34–3.45 g/cm³
Shale	2.06–2.66 g/cm³	Quartz	2.648 g/cm³
Limestone	2.37–2.80 g/cm³	Felspar	2.61–2.76 g/cm³
Mudstone	1.44–1.93 g/cm³	Calcite	2.712 g/cm³
Granite	2.52–2.81 g/cm³	Pyroxene	3.15–3.61 g/cm³
Amphibolite	2.79–3.14 g/cm³	Olivine	3.21–4.39 g/cm³

Let us now try to estimate the difference in hydrostatic pressure that will be produced by differences in the densities of the halite and the sedimentary rocks. If we assume that a salt formation, about 1500 m high, exists upwards from a depth of 8500 m the hydrostatic pressure at the bottom of the formation will be less than if it was not present, and there were sediments instead, with a density as shown in Figure 2.VIII.7. Thus the difference between these hydrostatic pressures, calculated from the density difference at a depth of 8500 m, will be $0.3 \times 150\,000 = 45\,000$ g/cm², or almost 45 bar. The relationship between the height of the salt column and the difference in the hydrostatic pressure at its base may be computed and yields the curve shown in Figure 2.VIII.7. These results demonstrate that as the salt layer is squeezed

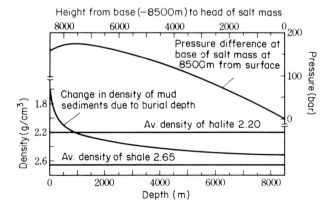

Figure 2.VIII.7   Changes in density due to the depth of burial of sediments and pressure difference at the base (−8500 m) of the salt dome (after Gussow, 1968*)

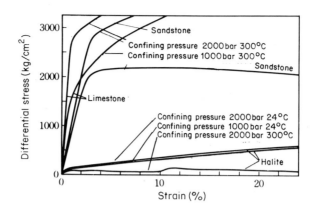

Figure 2.VIII.8   Deformation of halite, sandstone and limestone (Handin and Hager, 1958*)

up, it develops into a column and penetrates higher, so the difference in the effective hydrostatic pressure at its base increases, thus increasing the *buoyancy* effect.

At this point let us consider the results obtained by Handin and Hager (1958*) in their experiments on the deformation of rock salt. Salt is clearly deformed much more easily than sandstone or limestone, as shown in Figure 2.VIII.8. Its ultimate strength declines markedly with temperature and drops to about 100 bar at 300 °C with a confining pressure of 2000 bar. Moreover, rock salt can deform without fracturing. The experimental research of LeComte (1965*) into its creeping characteristics gave a creep curve for rock salt expressed by $\epsilon = A + Bt^n$. Taking constants determined by conditions such as temperature, strain rate and differential stress as A, $B > 0$, and expressing strain ϵ in per cent and time t in hours. The constant n will then be within the range $0 < n < 1.0$. For example, with a confining pressure of 1000 bar, differential stress of 69 bar and temperature of 104.5°C, $\epsilon = (64.0 + 5.54t^{0.622}) \times 10^{-4}$. One can conclude from these results that salt will readily flow in the upper crust. If it is permissible to regard the salt as a viscous fluid, it is possible to calculate its viscosity from the amount of deformation. Ode (1968*) obtained the viscosity, $\eta \cong 4 \times 10^{18}$ poise, from a rate of 0.9 mm/year for plastic deformation inside a salt tunnel, and he concluded, on the basis of many experiments and observations, that salt possesses a viscosity of 10^{17}–10^{19} poise.

(c)   Distribution of salt domes

A number of topographical elevations 25–40 km wide are known to exist on the sea floor of the Gulf of Mexico. Seismic surveys and drilling tests have shown the presence of salt domes inside each of these. Clastic sediments of the Tertiary to the Quaternary, composed principally of turbidites, are found around the salt domes. The sedimentary strata are inclined gently towards the outside of the dome structure and tend to become thinner towards the centre. Steep angled faults running almost parallel to the depth contours are found on the edge of the continental shelf. Some of the strata are cut by these faults and are successively downthrown towards the Gulf. These faults are thought to have been active during the late Tertiary and the Holocene. The edge of the continental slope, known as the Sigsbee Scarp, is said to be a submarine feature indicative of the presence of a huge salt wall, in which a number of intruding salt formations are joined together. Lehner (1969*) considered all these features together, and concluded that the salt was also flowing sideways in response to unequal loading, produced by the accumulation of terrestrial sediments. When one remembers the physical properties of rock salt and its overburden it is quite likely that plastic flow of the salt formation could occur not only in an upwards direction but also sideways. Structures (Figure 2.VIII.9) observed, from the southern part of the North American continent down into the Gulf of Mexico, all have their origins in the plastic flow of salt and are typical examples of structures resulting from salt tectonics.

The salt which creates the salt domes described above is believed to have a common source in a bed distributed widely and deeply below this region. Assuming it to have an average thickness of 300 m and an area of 5.5×10^5 km^2, the total volume of salt would be 165 000 km^3, and could well be even more than that (Murray, 1968*). Comparing it with the volume of Mount Fuji, i.e. about 400 km^3, illustrates just how prodigious this amount is. The salt is an *evaporite* deposit made by the evaporation of sea water. To make such a huge amount of salt would have required an unimaginable volume of sea water. Conditions in which neither mud nor sand became mixed up with the salt during the evaporation process must have prevailed for a long time

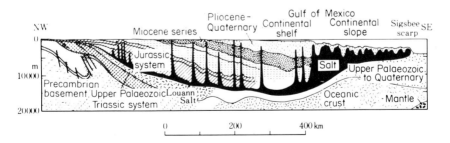

Figure 2.VIII.9 Geological cross-section in southern North American Continent (King, 1977*)

and over a wide area. Kirkland and Gerhard (1971*) investigated pollen grains obtained from a salt dome in the *Gulf of Mexico* and assessed the date of salt as Jurassic. At this time the whole of the Gulf of Mexico region was probably in an environment which would readily lead to the production of evaporites.

Let us look more closely at these environments in space and time. Evaporites, salt domes and their accompanying folded structures, are known not only on the North American continent but also throughout the continents of South America and Africa and on the ocean floor. As the presence of salt layers was verified and their stratigraphic relationships established, so geologists began to recognize the considerable similarities between them. Their distribution, as summarized in Figure 2.VIII.10, and the stratigraphy shown in Figure 2.VIII.11 are known from offshore Newfoundland to Morocco,

Figure 2.VIII.10   Distribution of sedimentary strata containing evaporites in Atlantic coastal areas (see Figure 2.VIII.11) (after Evans, 1978*)

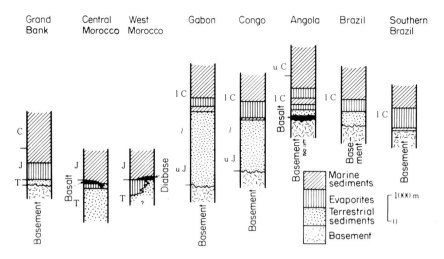

Figure 2.VIII.11   Stratigraphy of evaporites bearing formations in Atlantic coastal areas (see Figure 2.VIII.10)

Gabon and Southern Brazil. The age of salt tends to be Triassic in the more northern areas, whereas in the southern areas, it is rather younger, dating from the Cretaceous. These facts suggest that, as shown in Figure 2.VIII.10, the environment was suitable for the formation of evaporites at about the time when the continents, which had formed a unified structure, began to break away from each other, beginning with the development of huge fracture zones. Such movements in the Triassic led to the formation of structures resembling the present day African Rift System. Sea water then probably invaded the rift areas, evaporated and so produced the rock salt deposits.

Shinjiro Mizutani

References

Arrhenius, S. (1912): Zur Physik der Salzlagerstatten, *Meddel. Vet. Akad. Nobelinst.*, **2**, 1–25.

Atwater, G. I. and Forman, M. J. (1959): Nature of growth of southern Louisiana salt domes and its effect on petroleum accumulation, *Am. Assoc. Petrol. Geol. Bull.*, **43**, 2592–622.

Braunstein, J. and O'Brien, G. D., ed. (1968): Diapirism and diapirs, *Am. Assoc. Petrol. Geol. Mem.*, **8**, 444 pp.

Evans, R. (1978): Origin and significance of evaporites in basin around Atlantic margin, *Am. Assoc. Petrol. Geol. Bull.*, **62**, 223–34.

Gussow, W. C. (1968): Salt diapirism: importance of temperature, and energy source of emplacement, *Am. Assoc. Petrol. Geol. Mem.*, **8**, 16–52.

Handin, J. and Hager, R.V., Jr. (1958): Experimental deformation of sedimentary rocks under confining pressure: test high temperature, *Am. Assoc. Petrol. Geol. Bull.*, **42**, 2892–2934.

Kent, P. E. (1958): Recent studies of south Persian salt plugs (Iran), *Am. Assoc. Petrol. Geol. Bull.*, **42**, 2951–72.

King, P.B. (1977): *The Evolution of North America*, Revised ed., 197 pp., Princeton Univ. Press, Princeton.

Kirkland, D. W. and Gerhard, J. E. (1971): Jurassic Salt, central Gulf of Mexico, and its temporal relation to Circum-Gulf evaporites, *Am. Assoc. Petrol. Geol. Bull.*, **55**, 680–6.

Kupfer, D. H. (1962): Structure of Morton Salt Company mine, Weeks Island salt dome, Louisiana, *Am. Assoc. Petrol. Geol. Bull.*, **46**, 1460–7.

LeComte, P. (1965): Creep in rock salt, *J. Geol.*, **73**, 469–84.

Lehner, P. (1969): Salt tectonics and Pleistocene stratigraphy on continental slope of northern Gulf of Mexico, *Am. Assoc. Petrol. Geol. Bull.*, **53**, 2431–79.

Morgan, J. P. (1976): Louisiana deltaic geology, AAPG/SEPM Ann. Convention, Field Trip Guidebook, *Louisiana delta plain and salt domes*, 1–17.

Murray, G. E. (1968): Salt structures of Gulf of Mexico basin — a review, *Am. Assoc. Petrol. Geol. Mem.*, **8**, 99–121.

Ode, H. (1968): Review of mechanical properties of salt relating to salt-dome genesis, *Am. Assoc. Petrol. Geol. Mem.*, **8**, 53–78.

Spillers, J. P. (1969): *Salt Domes of South Louisiana*, Vol. 2, 107 pp., New Orleans Geol. Soc., New Orleans.

Trusheim, F. (1957): Über Halokinese und ihre Bedeutung für die strukturelle Entwicklung Norddeutschlands, *Z. deut. geol. Gesell.*, **109**, 111–58.

IX   *Sudbury astrobleme*

The mining town of Sudbury is located about 50 km north of Lake Huron in southeast Ontario, Canada. An oval-shaped igneous body is exposed to the north of the town, and has long been called the 'Sudbury Irruptive'. Many ore deposits of nickel and copper sulphides are found around the perimeter of the igneous body and the area has become the world's principal producer of nickel. Besides being of great economic significance, it is also of considerable geological and petrological interest, and has long been the subject of research. However, the geology of the area incorporating the *Sudbury Irruptive* shows some extremely unusual features which, despite painstaking research, still remain incompletely explained.

In the 1950s the attention of some geologists was drawn to structures created by the impact of meteorites. As coesite, a high pressure phase of SiO_2, was found around the Arizona Meteor Crater, and more information was amassed concerning fracture and metamorphic phenomena in rocks thought to have been produced by meteorite impact, not only meteorite impact craters leaving a topographical imprint, but also structures which were thought to have been produced by meteorite impact in ancient geological periods were identified one after another. An appropriate description for the latter is perhaps 'fossil' meteorite impact craters. Dietz (1960*) named such ancient scars produced by meteorite impact '*astroblemes*' from the Greek words for 'star' and 'wound'.

There are signs of magma outflows in what are presumed to be the huge meteorite impact craters on the moon. There are, however, virtually no such indications in impact craters on the earth. While conducting a survey to find areas on the earth equivalent to the seas of the moon, the idea struck Dietz that igneous activity in Sudbury provided a possible candidate. His survey of the area disclosed several structures he believed attributable to meteorite impact, on the basis of which he felt justified in explaining the entirety of Sudbury structures as a huge ancient impact crater (Dietz, 1964a*). These structures were subsequently studied by a number of geologists, whose works largely supported the fundamental proposition of Dietz, although they made many amendments. The idea that the Sudbury structure was an impact crater thus gained wide acceptance. Obviously the astrobleme hypothesis cannot explain every structure in Sudbury. Some people still saw it as a kind of volcanic structure. However, the accumulation of recent data has tended to lend more and more weight to the astrobleme concept. In this chapter, we present an outline of the geological, petrological and structural features found in the Sudbury area and consider to what extent they may be explained by the astrobleme hypothesis.

(a)   The Sudbury Irruptive

The Sudbury Irruptive is revealed as an ellipsoidal structure about 60 km along its long axis and 25 km along its short axis, lying in an ENE–WSW direction (Guy-Bray and Peredery, 1971*; Card *et al.*, 1972*). The igneous body, is 1.5 – 3 km thick (Figure 2.IX.1). As the shape of the structure resembles the hull of a boat, it has been called the 'Sudbury Basin', but the igneous body has also been called the 'Sudbury Lopolith' from its morphological features. Underneath the igneous body one finds metamorphic rocks derived from eruptive and clastic rocks together with granites and gneisses. It upper side is covered with layers of tuff, slates and sandstones known as the Whitewater series. A profile of the area shows at least part of the footwall rocks series to be steeply inclined forming a discordant structure with the igneous body. On the other hand, sediments on the hanging wall side apparently lie conformably over the igneous body in the basin structure. In short, the Sudbury structure may be seen as comprising three independent

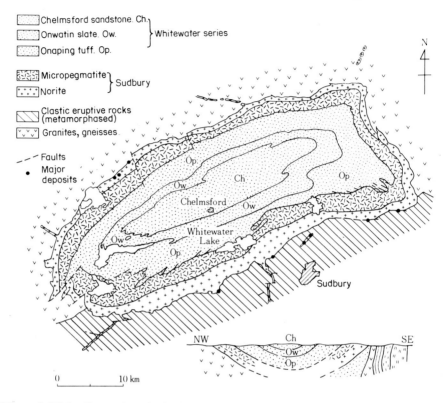

Figure 2.IX.1   General geological map and profile of the Sudbury area. The SE part of the profile shows the strata somewhat differently to the legend so as to clarify the structure

geologic units, i.e. (1) footwall rocks, (2) igneous body, (3) hanging wall rocks. Let us consider the features of each of these three units in turn:

(b)   Footwall rock — shatter cones and brecciation

The rocks underlying the Sudbury Irruptive, to the north of the basin-like stucture, are mainly Archaean granites and gneisses, but metamorphic rocks derived from eruptive and clastic material belonging to the Lower Proterozoic Huronian Supergroup lie to the south.

Strata on the southern side are often inclined steeply towards the south, and occasionally may even turn upwards or be locally overturned. Granitic rocks and gneisses predominate on the northern side, and their structure is more complex. It was felt that the basement to the Sudbury basin-like structure, when viewed as a whole, could perhaps be a single dome-shaped structure. Some profiles have been drawn, in which the Sudbury structure formed with the hollowing out of the head of the dome.

Two characteristic structures have developed in the rocks on the footwall, these are brecciation and the development of shatter cones.

Brecciated rocks occur in types of breccia dykes filling in fractures which had developed radially from the Sudbury basin or concentrically to it. Rather than being straightforward dykes, they consist of irregular patches or complex networks with a tremendous variation in scale. They are composed of blocks of various sizes, and matrix consisting of finely shattered rocks and glassy material. Most of the blocks clearly derive from the immediately surrounding rocks but there are also some extraneous types. Flow structures are observed in the matrix and give the impression of flow of somewhat viscous material.

Rock brecciation is most marked just at the base of the Sudbury Irruptive and dies out towards the perimeter but brecciation phenomena can still be seen more than 10 km away (Figure 2.IX.2). Such phenomena are not seen in the rocks of the Sudbury Irruptive itself or in the overlying Whitewater series.

Shatter cones are characteristic conical fractures which develop in rocks (Figure 2.IX.3). They are encountered quite commonly in the structures all over the world believed to be ancient impact craters as well as in the rocks surrounding young impact craters with topographic imprints. On the other hand, we have no certain knowledge of examples where such structures have formed following volcanic explosions. Hence, many people assume the cones

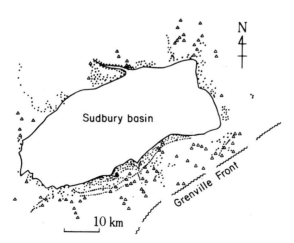

Figure 2.IX.2.   Distribution of Sudbury breccia and shatter cones (Dietz, 1964a, Guy-Bray *et al.*, 1960). ● Observed brecciated rocks. ▲ Observed shatter cones. Front of Grenville overthrust shown as 'Grenville Front'. The Grenville province exists to the SE of this line

Figure 2.IX.3   Shatter cones in Mississagi quartzite

to be fractures produced by the shock of meteorite impact. The height of the shatter cones varies from a few centimetres to over one metre.

When Dietz (1964a*) first conducted his survey little was known of shatter cones in the Sudbury area, but as detailed studies were undertaken it became clear that they had developed to a quite remarkable extent in rocks around the Irruptive (Dietz, 1964b; Guy-Bray *et al.*, 1966*.) Shatter cones may be found up to 13 km away from the edge of the Irruptive, but not in it or in rocks overlying it (Figure 2.IX.2). They are particularly well developed towards the south where the rocks are of a very compact sedimentary type known as Mississagi quartzite. This shows that the development of shatter cones may be facilitated or impeded by the properties of the rock.

The directions indicated by the points of shatter cones were also examined, but did not provide enough evidence to permit an accurate assessment of a point of origin of the shock waves. This was probably because it has been difficult to assess the effect of rock deformation which occurred after the shatter cones had been formed. However the apices of the shatter cones taken as a whole do tend towards a position a little higher than the current surface level in the centre of the Sudbury basin and does not contradict the idea of meteorite impact (Guy-Bray *et al.*, 1966*).

(c)   *The Sudbury Irruptive and ore deposits*

The upper half of the igneous body, which at first sight seems to be in the form of a lopolith is composed of granophyric rocks called micropegmatite, and the lower half of norite, i.e. gabbro with much orthopyroxene. They have sometimes been thought of as a type of ring dyke complex formed by the successive intrusion of two types of magma. Between the upper and lower rock units is a zone one may call a transition zone and the intrusive body as a whole can be explained as a differentiated layered intrusion. At the bottom of the hull-shaped igneous body there are discontinuus intrusive layers of norite, containing many xenolithic blocks of ultrabasic rocks and often being rich in sulphides. These are called the 'sub-layers'. Diorite similarly rich in xenoliths and sulphides also intrudes here and there into the footwall rocks sending off branches from the principal structures. These are referred to as 'offsets'.

According to some reports, Rb-Sr dating gives a value of 2000 Ma for norite in the main body and sub-layers, whereas the micropegmatite is put at 1700 Ma. However, the systematic variation in the mineral composition of the intrusives seems to show the rocks to be a single formation.

The rocks of the sub-layers and offsets, particularly rich in sulphides form the main nickel and copper sulphide deposits. Since the discovery of the first deposits in 1883, about fifty more have been found around the Sudbury Irruptive and many of these have been developed and exploited. Up till now some six million tons of nickel and an almost equivalent amount of copper have been produced.

The principal ore minerals are pyrrhotite, pentlandite, chalcopyrite, pyrite, cubanite and magnetite. Other minerals coexisting with the sulphide minerals include the same silicates, pyroxenes and plagioclases as those which make up the surrounding rocks. Phenomena such as hydrothermal alteration of the country rocks are not encountered. The mineral deposits are an intrinsic structural element of the Sudbury Irruptive and may be interpreted as one particular facies of norite or diorite rich in sulphides. Blocks of ultrabasic rock in the sub-layers and offsets incorporate material which would be expected to exist at depth as early products of differentiation.

At one time this area alone provided more than 70 per cent of the annual world production of nickel. The huge size of the deposits gave rise to speculation that the mechanism by which they had been formed perhaps differed from that of normal deposits. The results of extensive mineralogical research seem to indicate that they were produced from magma by an accumulation of immiscible sulphide liquid, and cannot be regarded as anything other than orthomagmatic deposits (Naldrett and Kullerud, 1967*; Souch *et al.*, 1969*).

(d)   Hanging wall rocks — Onaping tuff

Igneous rocks in the Sudbury Irruptive are exposed in a ring shape, with sedimentary material inside the ring being known as the Whitewater series. The Whitewater series is divided into three, with the Onaping tuff at the bottom overlaid by the Onwatin slate and the Chelmsford sandstone. The Onaping tuff is about 1500 m thick, and the whole Whitewater series itself is estimated to be more than 4000 m thick. Complex local folding and faulting are also visible and the whole structure is shaped like the bottom of a shallow boat. The Whitewater series is a specific set of strata found only on top of the Sudbury Irruptive. No corresponding strata are found outside the area.

At first sight the Onaping tuff appears as an accumulation of tuff and tuff breccia, containing many extraneous fragments. Deposits several tens of metres thick and composed of variously sized breccias are found at the very bottom of the Onaping tuff and the fragments often originate from the rocks surrounding the Sudbury basin structure. Sorting of the Onaping tuff is poor, with very little bedding. It also resembles thick welded tuff. Some estimates have put the total volume of the Onaping tuff at up to 1000 km^3, and it certainly gives an impression of the rapid deposition of a prodigious amount of extrusive material.

Chlorite and actinolite were formed in the rocks as result of weak metamorphism and recrystallization, but the basic structures of the rocks are well preserved, revealing splinters of volcanic glass with flow line structures. Upon microscopic examination, however, it was found that minute inclusions followed a specific face of the crystals of quartz fragments found in the tuff. They were found to run parallel to (0001) and ($10\bar{1}3$) of the quartz (French, 1967*). These faces correspond to the direction of lamella produced when quartz is subjected to *shock metamorphism*. Thus an idea was proposed that the Onaping tuff did not consist of simple volcanic deposits but might perhaps be regarded as deposits of rocks shattered and partially fused under the influence of shock waves produced by a meteorite impact. Micropegmatite, which forms the upper part of the Sudbury igneous body, has been found to intrude into the bottom of the Onaping tuff in places.

Graded bedding is well developed in the Onwatin slate and the Chelmsford sandstone overlying the Onaping tuff. There is also some alternation of thick layers of sand and thin layers of mud. These suggest that part of the sedimentary material was supplied and deposited as a result of turbidity currents. Some syndepositional deformation is apparent probably as a result of subaqueous sliding. To a certain extent they are reminiscent of a deep sedimentary basin. The sediments were supplied mainly from granite and quartzite of the surrounding basement. There is virtually no material supplied from the Onaping tuff.

(e) Interpretation based on the meteorite impact hypothesis

A summary of the Sudbury structures is given above. The hanging wall and footwall formations of the Sudbury Irruptive are composed of rocks which are totally different from each other in terms of both their lithology and structure. Moreover, strata of the hanging wall are found nowhere but actually inside the Sudbury basin and have not been identified anywhere beyond the perimeter. The lower part of the Whitewater series is composed of rocks resembling welded tuff. The Sudbury breccia, whose significance in the Sudbury structures had long been a riddle, develops only in the footwall rocks. These facts present many obstacles to regarding the Sudbury Irruptive simply as a lopolith intruding concordantly into the surrounding rocks. The discovery of shatter cones in the area as well led Dietz (1964a*) to the idea that the Sudbury structures had been formed as result of meteorite impact, i.e. to his 'astrobleme hypothesis'. According to this idea, the processes of formation of the Sudbury structures were as follows: (Figure 2.IX.4)

(i)   About 1700 million years ago a huge meteorite fell on to the Sudbury area, at great speed, gouging out a deep crater, about 45 km across. Shock waves spread throughout the surrounding rocks. The rocks became brecciated, shatter cones were formed and the rocks were subjected to shock metamorphism. Locally, rock was fused and spattered over the walls of the crater, and intruded into fractures to form the matrix of the breccias.

(ii)   As a result of the heat generation and pressure drop associated with the removal of some of the overburden, and triggered by the fall of the meteorite, magma formed under the ground, and welled up into the crater. The surface of the extruded magma cooled, forming a crust which further fractured and volcanic eruption occurred through the fractures. Thus rocks like thick welded tuff were formed on the surface of the magma pool. The interior cooled slowly and consolidated as a differentiated layered igneous body.

(iii) Sea water invaded into the crater, clastic material was provided from the surrounding area, and the Onwatin slate and Chelmsford sandstone which make up the sedimentary material of the Whitewater series were accumulated.

(iv)   About 1000 million years ago at the peak of Grenville orogeny the area was subjected to compressive forces from the southeast. As a result the circular Sudbury structure took on the oval shape seen today and gentle folding of the Whitewater series occurred. Locally, overthrust faults were formed.

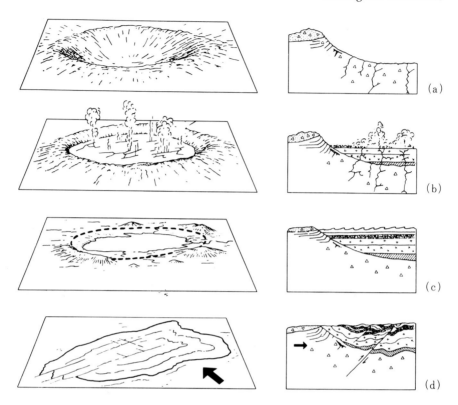

Figure 2.IX.4   Development of Sudbury Astrobleme (Dietz, 1964a*). (a) Formation of crater — surrounding strata upturned, rocks brecciated. (b) Crater becomes magma pool with ascended magma. Surface cools and forms crust which cracks allowing emission of more magma, building up rock like welded tuff. Interior of extruded magma cools slowly forming lopolith-like igneous body. (c) Deposition of clastic material in the Whitewater series. (d) Grenville overthrust and levelling with erosion

According to this hypothesis, the Sudbury Irruptive did not, in fact, intrude into the surrounding strata but should be called an 'extrusive lopolith' in which magma welled up into a meteorite impact crater and solidified in the form of what appears at first sight to be a lopolith. Almost no traces of shock metamorphism can be found, apart from the breccia and shatter cones, but the evidence may have been destroyed by subsequent regional metamorphism. The Sudbury basin structure was originally thought to have been formed over a basement with a dome-like structure, but it has been pointed out that upturning of the strata surrounding the crater as a result of the meteorite impact had probably been mistaken as a feature of dome structures.

The basic proposition of Dietz is endorsed by local mining geologists with detailed knowledge of the Sudbury area. However, there are inconsistencies in his explanation of the mechanism by which the Onaping tuff was created. French (1967*, 1972b*) later discovered planar structures in the quartz crystals of the Onaping tuff thought to be due to shock metamorphism, and suggested that the Onaping tuff was possibly deposited material, consisting of fused rocks and rock fragments scattered around by the impact and fragments which caved in from the walls of the crater. The identification of glassy material thought to have been formed from granitic rocks by the shock (Peredery, 1972*) gave rise to a belief that the Onaping tuff is not in fact a tuff but rather a peculiar clastic rock produced by the impact of the meteorite. With this concept the Sudbury Irruptive would become an igneous body which intruded between the basement rocks and the Onaping tuff and consolidated. This interpretation is thought to be quite feasible. However, what explanation can be found for the origins of the nickel and copper sulphide deposits around the Irruptive? Although he himself had misgivings about his interpretation, Dietz (1964a*) put forward the possibility that the nickel and copper sulphide deposits at Sudbury resulted from the descent of a Ni-Fe meteorite or one rich in sulphides. This melted and the sulphides permeated into the walls of the crater. Nowadays however, little weight is attached to this idea. Most research would seem to indicate that if magma had been generated by the meteorite impact, the deposits can be adequately explained by processes of differentiation within the magma. To sustain the idea of a Ni-Fe meteorite a mass of about 4 km in diameter would have been needed to create the Sudbury structure and in the case of less dense material, a still larger meteorite would have to be visualized.

A number of other meteorite impact craters or *astroblemes* are known throughout the world, but Sudbury is the only case where the impact of the meteorite is thought to have induced igneous activity (there is also an idea that the Bushveld intrusion in South Africa may be an astrobleme, but this is still uncertain). As was mentioned above, an interpretation is that a crater was made by the descent of a large meteorite and the consequent violent reduction in pressure brought about partial fusion of subterrranean material. It has been pointed out, however, that prerequisites to this would be a fairly large temperature gradient under the area at the time of meteorite impact and that conditions underground were favourable to the formation of magma (French, 1972a*). Is there any geological evidence in the area to suggest that such conditions were satisfied? There are a number of problems which require further investigation if the astrobleme hypothesis for the Sudbury structure is to gain more acceptance. However, it should be remembered that the introduction of a new approach, i.e. the astrobleme hypothesis, has promoted an extensive study and permitted a more unified interpretation of the Sudbury structure which had hitherto been regarded as exceedingly complex and miraculous in the conventional geological concept. The research history of the Sudbury structure does prove the importance of devising new ideas and approaches to controversial complex problems.

(f) Impact craters and fossil impact craters

Research into meteorite impact craters of fossil impact craters (astroblemes) has recently been making progress throughout the world, not just in Sudbury. Generally speaking, the following clues are looked for in the identification of a meteorite crater:

(1) The presence of meteorite material. Unless the crater is of recent origin this will not be found;
(2) A circular structure on the surface. However, subsequent crustal movements may have resulted in deformation;
(3) A structure in which strata surrounding the crater are turned up;
(4) The presence of mounds in the centre of the crater and the presence of brecciated deposits;
(5) Brecciation in the surrounding rocks;
(6) Changes in rocks due to shock metamorphism, e.g. development of shatter cones, presence of high density minerals, development of lamellar structures in minerals, vitrification of minerals. These features may disappear as a result of subsequent metamorphism;
(7) Anomalies in geophysical properties of the area, gravity, magnetism, seismic wave velocity and others.

In recent impact craters such as the Barringer Crater in Arizona which still remains topographically, much proof has been preserved, but this is far from being the case in fossil impact craters from ancient geological periods. The developments in research into shock metamorphism and the identification of shock metamorphic phenomena in many fossil impact craters have made the identification of several doubtful structures as meteorite impact craters much more reliable.

Dence (1972*) classified impact craters into three categories based on reliability of the available data, namely

(1) Certain impact craters in which meteorite material had been found;
(2) Probable impact craters in which shock-metamorphic features could be observed,
(3) Possible impact craters due to the circular shape of the structure, etc.

He cites 63 instances of (1), 42 of (2) and 39 of (3) from information for North America, Europe, Australia and parts of Asia and Africa. Representative examples of these are shown in Table 2.IX.1.

Table 2.IX.1 Examples of impact craters and fossil impact craters (after Dence, 1972*)

Name and Location	Latitude	Longitude	Diameter	No. of craters
			(m)	(150 m + diameter)
(1) Certain meteoric impact craters				
Aouelloul, Mauritania	20°15'N	012°41'W	250	1
Barringer, Arizona, USA	35 02 N	111 01 W	1200	1
Boxhole, Northern Territory, Australia	22 37 S	135 12 E	175	1
Henbury, Northern Territory, Australia	24 34 S	133 10 E	150	14
Odessa, Texas, USA	31 48 N	102 30 W	168	3
Wolf Creek, Western Australia	19 18 S	127 47 E	850	1

Largest one refers to the Diameter (m) column.

Name and Location	Latitude	Longitude	Diameter (km)	Age (Ma)
(2) Probable impact craters, structures with shock metamorphism (15 km + diameter)				
Carswell, Saskatchewan, Canada	58°27'N	109°30'W	30	485±50
Charlevoix, Quebec, Canada	47 32 N	070 18 W	35	350±25
Clearwater Lake East, Quebec, Canada	56 05 N	074 07 W	15	285±30
Clearwater Lake West, Quebec, Canada	56 13 N	074 30 W	30	285±30
Gosses Bluff, Northern Territory, Australia	23 48 S	132 18 E	22	130±6
Manicouagan, Quebec, Canada	51 23 N	068 42 W	65	210±4
Manson, Iowa, USA	45 35 N	094 31 W	30	Mesozoic
Mistastin, Labrador, Canada	55 53 N	063 18 W	20	202±25
Ries, Germany	48 53 N	010 37 E	24	14.8±0.7
Rochechouart, France	45 50 N	000 56 E	15	165±10
St Martin, Manitoba, Canada	51 47 N	098 33 W	24	225±25
Siljan, Sweden	61 05 N	015 00 E	45	Palaeozoic–Mesozoic
Steen River, Alberta, Canada	59 31 N	117 38 W	25	95±7
Strangways, Northern Territory, Australia	15 12 S	133 35 E	16	Palaeozoic–Mesozoic
Sudbury, Ontario, Canada	46 36 N	081 11 W	100	1700±200
Vredefort, South Africa	27 00 S	027 30 E	100	1970±100

Name and Location	Latitude	Longitude	Diameter (km)	
(3) Possible impact craters (15 km + diameter)				
Haughton Dome, Northwest Territories, Canada	75°22'N	089°40'W	17	–
Labynkyr, Yakut, USSR	62 30 N	143 00 E	60	–
Popigay, Taymirskiy-Yakut, USSR	71 30 N	111 00 E	65	–
Puchezh-Katun,	57 06 N	043 35 E	70	–
Zhamanshin, Aktyubinsk, USSR	49 N	059 E	15	–

Much exploration for and into impact craters and fossil impact craters has taken place in the *Canadian Shield*. It is now believed that in that area almost all the impact craters (mainly fossil impact craters) 20 km or more in diameter which are capable of being identified in their circular structures have been found. The ages of impact craters have also been the subject of intense study using radiometric methods. As a result a geographical and temporal distribution of impact craters has been established and we are now able to estimate with much greater accuracy than before the frequency with which meteorite impact craters are formed.

The geographical distribution of impact craters is not uniform even in the closely studied Canadian Shield, nor is the frequency of crater formation uniform throughout the geological periods. There appears to be a particularly large number of impact craters in areas such as Quebec–Labrador. Apart perhaps from chance, this is thought to be due to regional differences in erosion (Dence, 1972*).

Looking at the Canadian Shield as a whole, it is estimated that one meteorite impact crater more than 20 km in diameter will be formed in an area of 10^6 km^2 in 10^9 years. This value increases approximately fivefold for the Quebec–Labrador region where more impact craters are found (Dence, 1972*). Looking at this latter value for a moment, it corresponds roughly with the frequency of craters which are preserved so well on the moon, on the side facing the earth.

The distribution of impact craters or fossil impact craters so far identified on the earth is very uneven. Quite obviously, whether the preservation of a crater is good or bad will be considerably affected by the extent of subsequent crustal movement in the relevant area, but it is thought that the future research will establish their geographical and temporal distributions with much more certainty.

Keiichiro Kanehira

References

Card, K. D., Naldrett, A. J., Guy-Bray, J. V., Pattison, E. F., Phipps, D. and Robertson, J. A. (1972): *General geology of the Sudbury Elliot Lake region*, Excursion C 38 Guidebook, XXIV IGC, 1972 Montreal, Quebec, 56 pp.

Dence, M. R. (1972): The Nature and significance of terrestrial impact structures, XXIV IGC, Sect. 15, Planetology, 77–89.

Dietz, R. S. (1960): Meteorite impact suggested by shatter cones in rocks, *Science*, **131**, 1781–4.

Dietz, R. S. (1964a): Sudbury structure as an astrobleme, *J. Geol.*, **72**, 412–34.

Dietz, R. S. (1964b): Shatter-cone orientation at Sudbury, Ontario, *Nature*, **204**, 280–1.

French, B. M. (1967): Sudbury structure, Ontario: Some petrographic evidence for origin by meteorite impact, *Science*, **156**, 1094–8.

French, B. M. (1972a): Production of deep melting by large meteorite impacts: The Sudbury structure, Canada, XXIV IGC, Sect. 15, Planetology, 125–132.

French, B. M. (1972b): Shock-metamorphic features in the Sudbury structure, Ontario, a review, *Geol. Assoc. Canada, Spec. Paper*, No. 10, 19–28.

Guy-Bray, J. and Geological Staff (1966): Shatter cones at Sudbury, *J. Geol.*, **74**, 243–5.

Guy-Bray, J. and Peredery, W. V. (1971): Field guide for excursion (1) Sudbury basin, *Geol. Assoc. Canada Miner. Assoc. Canada*, 18 pp.

Naldrett, A. J. and Kullerud, G. (1967): A study of the Strathcona mine and its bearing on the origin of the nickel-copper ores of the Sudbury district, Ontario, *J. Petrology*, **8**, 453–531.

Peredery, W. V. (1972): Chemistry of fluidal glasses and melt bodies in the Onaping formation, *Geol. Assoc. Canada, Spec. Paper*, No. 10, 49–59.

Souch, B. E., Podolsky, T. and Geological Staff the International Nickel Co. of Canada, Ltd. (1969): The sulfide ores of Sudbury: Their particular relationship to a distinctive inclusion-bearing facies of the Nickel Irruptive, *Econ. Geol., Mon.*, **4**, 252–61.

Geological Structures
Edited by T. Uemura and S. Mizutani
©1984 John Wiley & Sons Ltd.

3

Joints, *Faults* and Stress Fields

The internal structure of the rocks and strata, which make up the earth's crust, frequently contains areas disturbed in various ways. Such structural disturbances include those of a primary nature, formed when the rocks and strata were themselves laid down, and also secondary types, created by subsequent tectonic processes. The latter may further be divided into continuous disturbances, such as folding, and disturbances due to various types of discrete surfaces such as faults and joints. Secondary discrete surfaces seen in geological bodies are generally known as 'fractures' or 'fracture planes'. There are many kinds of fractures such as joints, faults, cleavages and so on, and these may be classified still more finely in terms of their origin, shape and development.

Phenomena in which a body is divided into two or more parts by stress may be called '*failure*' and the ability to deform without fracturing '*ductility*'. Fracturing is failure in the range of relatively low ductility, but as ductility increases there is a transition towards continuous deformation (flow) via the formation of so-called foliated structures such as cleavage or schistosity. Fracturing is a common structural element in most geological bodies regardless of time, location, size or type. An understanding of fracture mechanisms is thus of overriding significance to any discussion of problems concerned with geological structures. Furthermore, a fracture plane once formed, shows strong anisotropy as a plane of weakness and may become the site of a dyke or the accumulation of mineral deposits. A pre-existing fracture, being a plane of separation within the rock, may also become a primary factor in slope failure such as landslides or landslips. They also pose major problems in the geotechnical assessment of bedrock for the construction of roads, bridges, tunnels, railways, and the foundations of buildings.

When an external force is applied to a solid body, a specific stress state is produced inside it. If the deformation produced as a result exceeds certain limits the body will fracture. In such a case one may identify the result as a *brittle fracture* or a *ductile fracture* from the form of the stress–strain curve and the position of the fracture point on the curve. The characteristics of the stress, causing the fracture, may also be used to identify *tension fractures* and *shear fractures*. Griggs and Handin (1960*) applied the terms '*extension* *fracture*' and '*fault*' to events corresponding to tension fractures and shear fractures respectively. Because shearing involves some displacement along the shear plane, no matter how little, their idea was probably that a shear fracture must be a fault from the beginning of fracturing. In general however, it is possible to distinguish between fractures involving a simple loss of cohesion and faults based on progressive displacement. By tension fractures we mean those produced in a stress state of absolute tension (i.e. when the principal compressive stress is positive and taking σ_1, σ_2, σ_3 from the largest, $\sigma_1 > \sigma_2 > 0 \gg \sigma_3$). However, because the same kind of result may be produced under the stress state of relative tension (i.e. $\sigma_1 > \sigma_2 \gg \sigma_3 > 0$) both may be considered as extension fractures, but only the former is called a tension fracture (Nagumo, 1968). This chapter mainly looks at the problems of stress fields which form faults or joints. It should be emphasized that in the majority of cases, even if the stress field can be determined from the fracture, the external force cannot be uniquely determined from the field. Commonly a number of possible mathematical solutions may be obtained for the external force. In such cases geological structures in the whole area, containing the joints and faults in question, must be investigated to determine their tectonic history. Not until the series of external forces most in accord with this are identified can one hope to arrive at the accurate solution.

3.1 Mechanical conditions of fractures

Numerous proposals have been put forward for the mechanical criteria related to the fracture of solid materials. In this section we look at the relatively simple ones applicable to shear fracturing in rocks and strata.

Mohr (1900) found that when a solid fractures there is a certain relationship $|\tau| = f(\sigma)$ between a normal stress σ and a shear stress τ acting on the fracture plane. On the other hand, if an ordinary three-dimensional stress state is expressed by principal stresses $\sigma_1 > \sigma_2 > \sigma_3$, then σ and τ for an arbitrary plane making at an angle θ with the σ_1 axis will be, as is well known:

$$\sigma = \frac{\sigma_1 + \sigma_3}{2} - \frac{\sigma_1 - \sigma_3}{2} \cos 2\theta$$

$$\tau = \frac{\sigma_1 - \sigma_3}{2} \sin 2\theta$$

These two equations form a circular equation expressed by:

$$\tau^2 + \left(\sigma - \frac{\sigma_1 + \sigma_3}{2}\right)^2 = \left(\frac{\sigma_1 - \sigma_3}{2}\right)^2$$

Putting this on to a rectangular co-ordinate plane with τ and σ on the axes gives us Mohr's stress circle. The values of τ and σ for an arbitrary θ may be given by co-ordinates on the stress circle. Therefore fracturing is expressed as a point of contact between the curve expressed by $|\tau| = f(\sigma)$ and Mohr's stress circle which shows the stress state applied. If fracture experiments are performed on the same material in various stress states and the co-ordinates of fracture points obtained on these stress circles are joined up, this should express a curve $|\tau| = f(\sigma)$ for the failure criterion for that material. As this gives an enveloping curve around the stress circles, the criterion is known as the *stress circle envelope theory* or *Mohr's theory* (Figure 3.1).

Two important criteria of failure are unified by Mohr's theory. The first is called *Coulomb's* criterion or the theory of the angle of internal friction. Follow compression tests with rocks and soil, Coulomb (1773) discovered that the relationship between τ and σ at the time of fracturing can be approximated with a straight line in which $|\tau| = \tau_o + \sigma \tan \psi$ or taking $\psi = \mu$ $|\tau| = \tau_o + \mu\sigma$ (Figure 3.2). τ_o and ψ or μ are known as the *cohesion* and *angle of internal friction* or coefficient of internal friction respectively, and are specific constants for each material. Remembering that r and 0 constitute Mohr's stress circle,

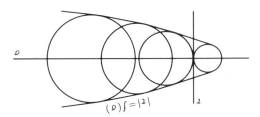

Figure 3.1   Stress circle envelope theory

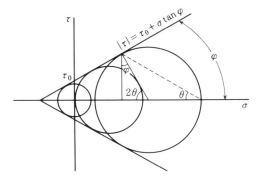

Figure 3.2   Theory of angle of internal friction

this condition represents the case in which Mohr's stress circle envelope is expressed by the straight line

$$|\tau| = \tau_o + \sigma \tan \psi$$

The second is Tresca's criterion (1864) or theory of maximum shear stress. Tresca conducted experiments into yielding conditions for metals and came to the conclusion that fracture and yielding occur when the *maximum shear stress* reaches a specific value for that material. Because the maximum value of the shear stress τ is $(\sigma_1 - \sigma_2)/2$, and bearing in mind its relationship with Mohr's stress circle, this represents the case when the stress circle envelope given in Mohr's theory will be a straight line parallel to the 0 axis given by $|\tau| = (\sigma_1 - \sigma_3)/ = $ constant (Figure 3.3). In Coulomb's criterion this may be regarded as the case when $\psi = 0$. In short, these two hypotheses are both special cases of Mohr's theory.

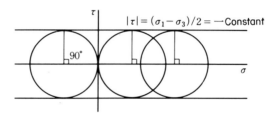

Figure 3.3   Theory of maximum shear stress

The idea common to the failure criteria in these and Mohr's theory is that they try to express $|\tau| = f(\sigma)$ within the parameters of specific constants for a particular material, i.e. 'physical' constants. In contrast, the 'field' of stress, expressed by equations with τ and σ, are unrelated to the physical properties of the material to which it is applied. In other words, when curves or straight lines, which express the 'physcial properties' relating to fracturing, and stress circles expressing 'field' are in contact, the phenomenon of fracturing occurs and the physical manifestation of this contact is the concept of the envelope curve. A line expressing this type of failure criterion is known as a 'fracture line'.

A number of theories exist concerned with the mechanical conditions of fracturing but there are currently few instances when they can be applied to a geological study of fractures. We do not therefore consider them here but discuss Coulomb's theory in a little more detail. The theory is somewhat irrational in the tensional field, and real fracture lines in rocks have been judged experimentally to resemble parabolas rather than straight lines.

However, the theory is still widely applied to problems of fracturing because the analysis is extremely easy, in addition to being able to obtain a reasonable degree of similarity with straight fracture lines.

Hubbert (1951*) combined shear stress $\tau = \sin 2\theta . (\sigma_1 \sigma_3)/2$ with Coulomb's failure criterion $|\tau| = \tau_o + \sigma \tan \psi$ to illustrate shear fracturing (Figure 3.2). Shear stress can take a maximum value when $\sin 2\theta = 1$, i.e. $\theta = 45°$. If one assumes a shear fracture to originate along the plane of maximum shear stress, then there should be a pair of planes inclined at 45° to the axis. However, this is a conclusion from the stress field only without taking the physical properties into consideraton. (The same results are obtained with the Tresca's theory taking $\psi = 0$.) Because there is an *angle of internal friction*, a plane of shear fracture will develop in the direction $2\theta = 90° - \psi$, i.e. $\theta = 45° - \psi/2$ as is clear from Figure 3.2. Because the value of τ is unrelated to σ_2 in this case, the shear plane should run parallel to the σ_2 axis. Furthermore, the experimental work of Mogi (1967) demonstrated that the effects of σ_2 cannot be ignored in the case of tension fracturing under confining pressure.

3.2   Restortion of stress fields

(a)   The state of crustal stress and classification of faults

Before considering the problems of stress fields, which give rise to fracturing, let us first examine the state of stresses present in the crust. Anderson (1951*) thought that a hydrostatic state of stress was maintained in the crust when not subjected to additional tectonic stresses and called this the '*standard state of stress*'. Therefore if one uses rectangular co-ordinates with the vertical direction as the z axis, the state of stress at a depth *z* may be written as:

$$\sigma_x = \sigma_y = \sigma_z = \rho g z$$

ρ is the average density of crustal material above this position and g is acceleration due to gravity. In contrast to this idea, which assumes crustal material to be in a state of fluid equilibrium, Price (1959*) thought it more appropriate to see it as an elastic body. Taking the same co-ordinate axes as above and expressing Young's modulus for crustal material as E and the Poisson's number as m, then strain in the horizontal direction may be determined as follows, from Hooke's law:

$$\epsilon_x = \frac{1}{E}\left\{\sigma_x - \frac{1}{m}(\sigma_y + \sigma_z)\right\}, \qquad \epsilon_y = \frac{1}{E}\left\{\sigma_y - \frac{1}{m}(\sigma_x + \sigma_z)\right\}$$

In general, $\epsilon_x = \epsilon_y = 0$, because of constriction by the surrounding rocks, if these are imagined to be continuous bodies without gaps. Therefore:

$$\sigma_x = \sigma_y = \frac{\sigma_z}{m-1} = \frac{\rho g z}{m-1}$$

m will decrease as the overburden pressure increases, but will generally be greater than 2, and σ_x and σ_y are lower than under Anderson's standard state of stress. This is called Price's standard state. Perhaps conditions in the real crust approach the Price standard state at shallow depths, but are closer to the Anderson state at greater depths, as the temperature and pressure increase. Examples of the values of stresses at work in the crust today, are detailed in Section 6.3.

Stress fields, giving rise to fracturing, appear when stress deviates from the standard state and loses its equilibrium. Anderson (1951*) envisaged directions for the principal stress axes when this occurred, as follows: because the surface of the earth is a free surface and shear stress along it is zero, one of the axes of principal stress at or near the surface can be thought to be vertical. The other two axes will, therefore, be horizontal, and there are only the three arrangements for the principal axes stress shown in Figure 3.4(a). If

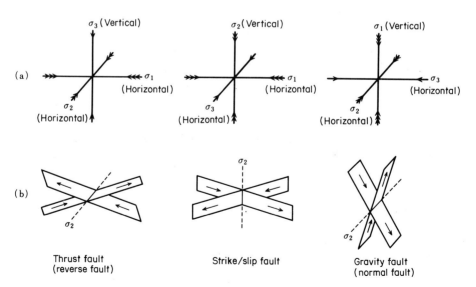

Figure 3.4   Arrangement of principal stress axes and fault types (after Anderson, 1951*)

Coulomb's failure criterion is applied to these three cases, the types of shear fracture, leading to fault formation illustrated in Figure 3.4(b) may be expected. Anderson distinguished faults formed in the crust, by linking them to these three arrangements for the principal stress axes and used them for a mechanical classification of faults.

(b) Conjugate faults and principal axes of stress

According to Coulomb's failure criterion, the relationship between the directions of the shear fracture plane and the principal stress axes will be expressed such that 'the shear fracture plane is parallel to the intermediate principal stress axis, and will develop in a direction making an angle of 45° − $\psi/2$ with the axis of maximum compressive stress'. As this direction can exist quite equally on both sides of the axis of maximum principal compressive stress, this is known as a set of *conjugate shear planes*. The angle, 2θ, formed between the two planes gives the *angle of shear* ψ from $2\theta = 90 \psi/2$ (Section 3.1). Because the faults which develop from them are *conjugate faults*, if they can be identified it should be possible to determine the direction of the principal stress axes. Kakimi (1968[*]) put forward three criteria for conjugacy in faults, namely the sense of the movement (as in Figure 3.4(b) if conjugate), severance relationships (contemporaneous formation if they cut each other) and other properties of the fault planes (similar if conjugate). He emphasized that all three conditions must be satisfied for faults to be conjugate.

Because active faults and particularly earthquake faults may largely be considered to be contemporaneous in geological terms, they may reasonably be regarded as conjugate faults, if conjugacy is present in the sense of their movement. Sugimura and Matsuda (1965[*]) and Huzita (1969[*]) regarded active fault groups in the central Japan and the Kinki region as conjugate faults, composed of left lateral faults running NW–SE and right lateral faults running ENE–WSW. They inferred that the principal axis of maximum compressive stress was lying in an almost E–W horizontal direction (Figure 3.5). This conclusion corresponds well with the *in situ* stresses measured in the bedrock at the Tokyo Denryoku Takasegawa underground power station. These give σ_1 as being inclined at 14–15° to the west and σ_3 15–16° to the north (Mimaki, 1973; Hiramatsu *et al.*, 1973).

(c) Minor faults analysis

It is generally desirable to observe both faults of a conjugate set in one outcrop in order to identify the stress field at the time of faulting. So-called 'minor faults' with only small displacements along the fault planes are convenient for this. Because many more minor faults exist than major ones, the direction of the principal stress axes can be determined at many localities

Figure 3.5   Stress trajectories in south-west Japan in early middle Pleistocene obtained from active faults and anticline elevations (thick lines are their axes) (Huzita, 1969*). Broken lines are trajectories of maximum compressive stress

and so they are convenient in statistical analysis, when attempting to average out the scatter in measured values, due to the differing properties of the rock.

It is useful to use a stereonet, such as shown in Figure 3.6, to determine the directions of the principal stress axes in order to restore palaeo-stress fields, from the numerous measurements of minor conjugate faults. If one draws great circles A and B for the average direction of the conjugate faults (P_A and P_B being their poles), the line of intersection gives the intermediate principal

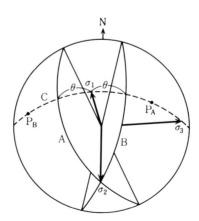

Figure 3.6   Stereographic projection and principal stress axes of conjugate faults

stress axis σ_2. By drawing in the great circle C with σ_2 as its pole and finding its intersection with the great circles A and B, the direction of σ_1 can be obtained as the bisectrix of the acute angle 2θ, and of σ_3 as the bisectrix of the obtuse angle.

If the conjugate minor faults, revealed in individual outcrops, were formed contemporaneously, the directions of the principal stress axes, thus obtained, may be combined to restore the regional palaeo-stress fields. Restored stress fields may be represented by inscribing principal stress lines (or stress trajectories) on to the plan and profile. If other conjugate fault systems are present it is possible to identify, approximately their age relationships from their cross-cutting relationships and to deduce changes of the palaeo-stress field with time.

The first person to explain regional stress fields by actually applying minor fault analysis in the field was Gzovsky (1954*). Much research has also been undertaken in Japan since the work of Fujita *et al.* (1965) and Hirayama and Kakimi (1965). The following example shows the application of the analysis to folding mechanisms in the Cenozoic formations in the Niigata region of central Japan, namely buckling (due to lateral compression) and transverse bending (due to elevation of the basement blocks). Uemura and Shimohata (1972*) analysed stress fields inside folded structures by applying the techniques of minor fault analysis to folds of the Plio-Pleistocene Uonuma group around the River Shimbumigawa Valley in central Niigata. They concluded that there is a zone of compression in the inner arc of folds and tension in the outer arc, with a *neutral surface* between the two (Figure 3.7), and thought this showed that buckling due to lateral compression must be considered in any interpretation of the mechanisms of these folds.

Another example of the analysis of folding mechanisms is the case of the *Niitsu anticline*. (Niitsu Anticline Research Group, 1977*). Numerous minor faults are developed along the Niitsu anticlines which are composed of Neogene strata, one typical example being the so-called 'oil field folding'. Amongst these, it was discovered that the directions of the principal stress axes obtained from minor faults (main fault system) were at right angles to the bedding planes (σ_1), parallel to the anticlinal axis (σ_2) and at right angles to the anticline axis whilst being parallel to the bedding plane (σ_3) (Figure 3.8). This regularity did not vary, despite variations in the direction of the anticlinal axis and no minor fault system showing horizontal compression could be found in this area. These two facts suggest that the Niitsu anticline was formed as a result of transverse bending of the Neogene sedimentary cover, accompanying upheaval of the basement.

(d) Problems of minor faults analysis

Because of the extreme rarity of natural *strain indicators* such as deformed pebbles or fossils, minor folds or boudins in the majority of younger strata,

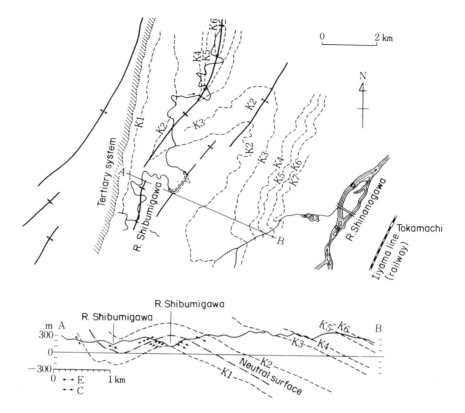

Figure 3.7 Folding in the Uonuma group (Uemura and Shimohata, 1972*) K1 – K7 key beds. Arrows E and C show directions of axes of minimum, and maximum principal compressive stress, respectively

almost the only means of assessing the stress field is minor faults. Mechanical analysis of geological structures has made great progress as a result of the application of minor fault analysis, but it is not without problems.

The first problem in reconstructing palaeo-stress fields by minor fault analysis is in its qualitative nature. The fact that the 'stress field' obtained in this way is expressed solely in terms of the directions of the principal stress axes means that they can only be considered in a qualitative fashion. Not until absolute stress values are calculated in such stress fields can the complete stress field be reconstructed. One must therefore determine the cohesion τ_o and the angle of internal friction ψ, etc. by triaxial tests on the rocks and must consider the relationship of these to their shear strength. When determining such physical constants, one difficult problem encountered in the assessment of the events of a geological period, is that values obtained from samples collected now will not be the same as the properties which existed when the fault was formed. This does not pose a problem when consolidation of the strata is regarded as having advanced to any great extent since the time when the fault was formed. It can probably also be disregarded in the case of igneous rocks. However, the relationship between the age and the failure strength of strata must be investigated in most sedimentary rocks and the measurements must be corrected as appropriate. One example of this is touched upon in Section 6.4 'The formation of boudins'.

The second problem is that of *mechanical anisotropy*. The material to be ruptured is assumed to be uniform and isotropic with respect to the mechanical failure criterion. In nature, the majority of rocks and strata are certainly neither uniform nor isotropic. This means that the constants τ_0 and ψ contained in the Coulomb's equation will vary with direction, and will affect the relationships between the directions of the principal stress axes and the attitude of the fault plane. Jaeger (1960*) considered this problem theoretically, as planar anisotropy in rocks with planar structures with minimal shear

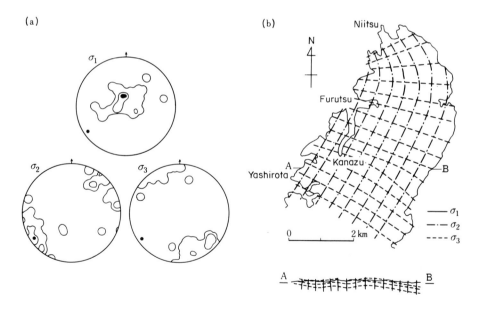

Figure 3.8   (a) Stereographic projection of principal stress axes in the central area of Niitsu anticline obtained from minor faults. Solid circles are the anticlinal axis. (b) Stress trajectories in Niitsu anticline obtained from main fault system (Niitsu Anticline Research Group, 1977*)

strength (weak planes). He found that a shear fracture plane originates between the plane of anisotropy and the plane, which is nearer the anisotropic plane, of the pair of conjugate shear planes which should occur in an isotropic body. He verified this further with fracture tests using slate. If planar structures, such as bedding planes, are present, single-directional faults may occur and not conjugate faults. The direction of these faults will also cease to correspond to that expected in an isotropic body. Donath (1961*) performed a series of triaxial tests using the Ordovician *Martinsburg slate* to examine the effects of anisotropy on the shear angle, when the angle formed between the loading axis, i.e. the σ_1 axis, and the slaty cleavage plane was altered. His results supported the ideas of Jaeger. Jaeger's concept was based on the hypothesis that τ_o in Coulomb's criterion $|\tau| = \tau_o + \sigma \tan \psi$ is not a constant but varies continuously in the relationship $\tau_o = a - b \cos 2 (\alpha - \beta)$. α and β are respectively the angles between the σ_1 axis and an arbitrary plane and an anisotropic plane. a and b are specific constants derived from the physical properties of the material.

When the effects of anisotropy are considered in this way considerable differences may well be produced between the direction for the principal stress axes obtained from conjugate faults and reality. Minor faults analysis must therefore be examined with great caution. This is particularly true when restoring stress fields from conjugate fault groupings in which one directional system predominates.

Yet another problem lies in the fact that stress fields obtained from minor faults do not inevitably express the regional stress field. When faults are formed or earthquakes occur in the regional stress field, the resulting elastic waves may set up a secondary stress field leading to the formation of secondary minor faults. As the direction of such a secondary stress field will differ from that of the original stress field any interpretation requires considerable care (Uemura, 1976*).

3.3   The form of stress fields

As discussed earlier, it is possible to obtain a stress field from a structure but the external forces at work cannot be determined uniquely from the stress field. To solve this problem, methods are adopted which give geologically feasible external conditions as external forces or limiting values of stress and displacement. Thereby, a theoretical stress distribution for the interior of the structural body in question is obtained, and the structures produced can be contrasted with those occurring in nature. Because geological deformation is extremely slow, when considering geological bodies of a specific size and shape, the boundary stress and the internal stress may usually be regarded as being in balance. *Airy's stress function* has been used in cases when an elastic body is taken to be in such a state of static equilibrium, so as to determine the internal stress distribution. Together with the advances in computer science, the finite element method has come to be the standard tool for such objectives. However, analytical methods based on stress functions together with stream functions also played a major role in mechanical analyses of geological structures throughout the 1950s and the 1960s and so we would like to give a broad explanation of such methods in this chapter. For more detailed treatments, the reader should refer to textbooks on elastic theory.

For a certain stress system which can exist inside an elastic body, one must verify that the stress system satisfied several equations demanded by elastic theory. This problem is solved, however, if only a single stress function is satisfied.

For the sake of simplicity let us consider a two-dimensional problem relating to the state of planar stress in the plane x–y. In this instance σ_x, σ_y and τ_{xy} are the only stress components acting on a single point. The stress function ϕ may be defined as follows

$$\sigma_x = \frac{\partial^2 \phi}{\partial y^2}, \qquad \sigma_y = \frac{\partial^2 \phi}{\partial x^2}, \qquad \tau_{xy} = -\frac{\partial^2 \phi}{\partial x \partial y}$$

Therefore, if the function ϕ is known, the stress components may be obtained immediately from the above definitions and it is also possible to obtain the principal stresses σ_1 and σ_2 and the angle β which gives their directions (β: the angle between the direction of σ_x, σ_y and the direction of σ_1, σ_2), and the maximum shearing stress $\tau_{\max}$ from the following well-known equations

$$\sigma_1, \sigma_2 = \frac{\sigma_x + \sigma_y}{2} \pm \tau_{\max}, \quad \tan 2\beta = \frac{2\tau_{xy}}{\sigma_x - \sigma_y}, \quad \tau_{max} = \sqrt{\left(\frac{\sigma_x - \sigma_y}{2}\right)^2 + \tau_{xy}}$$

Moreover, there are three essential conditions which the stress function must satisfy, namely the equilibrium equation of stress, Hooke's law (stress–strain relationship) and the compatibility equation for strain. Introducing these conditions and rearranging the equation gives:

$$\frac{\partial^4 \phi}{\partial x^4} + 2\frac{\partial^4 \phi}{\partial x^2 \partial y^2} + \frac{\partial^4 \phi}{\partial y^4} = 0$$

Using the Laplacian operator this may be written simply as:

$$\nabla^2 (\nabla^2 \phi) = \nabla^4 \phi = 0$$

In brief, solutions for all the problems relating to stress distribution in an elastic body can be obtained by a stress function ϕ which satisfies this biharmonic equation. Functions are of various forms in practice, including polynomial, exponential and periodical functions, etc. Below we give a few examples of their application to the problems of geological fracturing.

(a)   Stress fields and fracturing due to horizontal compression

Hafner (*1951**) investigated stress distributions inside bodies using stress functions, taking the various types of horizontal compressive stress or shear expected in the earth's crust as boundary conditions for the bodies. He predicted the type and fracture patterns of faults in the crust from his results. Taking the z axis as the vertical direction and the x and y axes in the horizontal direction, and further assuming the boundary conditions at the earth's surface to be $\tau_{2x} = \tau_{2y} = 0_1$ $\sigma_2 = 0$ and the vertical stress due to overburden as $\sigma_2 = -\rho g 2$ (ρ = average density), he first considered the case in which the stress field inside the crust is composed of hydrostatic stress and horizontal tectonic stresses applied to it. Two instances of these are shown in Figure 3.9(a) and (b). The first shows the case when $\sigma_x = ax - \rho g 2$ (horizontal compressive stress, equals the sum of the tectonic stress, proportional to the horizontal distance, and the hydrostatic stress which is proportional to the depth), $\sigma_z = -\rho g 2$ (vertical stress is equal to hydrostatic stress and proportional to depth) and $\tau_{xz} = -az$ (shear stress is porportional to depth). The second shows the case when σ_x tectonic stress is proportional to depth. Lines of equal maximum shear stress are expressed as multiples of a constant as in Figure 3.9, and shear fracture planes and faults arising from them are drawn in on the basis that $\theta = 30°$. The value of ρ/a is taken as the criterion for whether fracturing will occur or not and the cases are shown when this ratio is 1.0 and 2.0 as the boundary of the stable zone. This is a ratio between the pressure gradient in the vertical direction due to gravity and the lateral gradient of the applied horizontal stress. As can be judged from the figure, when its value is about 2 the area of *overthrust faulting* is limited to shallow parts but as a increases, the thrust zone develops at greater depths. When one also remembers that the conjugate overthrust faults shown by broken lines in the figures underneath (a) and (b) are very rarely found in nature, this pattern has a strong resemblance to folded zones accompanying single-directional low angle thrust faulting, and is useful to a consideration of its stress field. Hafner investigated various other cases besides this.

(b)   Stress fields and fracturing due to vertical displacement

Whereas *Hafner* gave all boundary conditions in terms of stress, *Sanford* (*1959**) analysed stress fields and fracturing which occur inside an elastic body when vertical displacement is applied to the base of the isotropic uniform

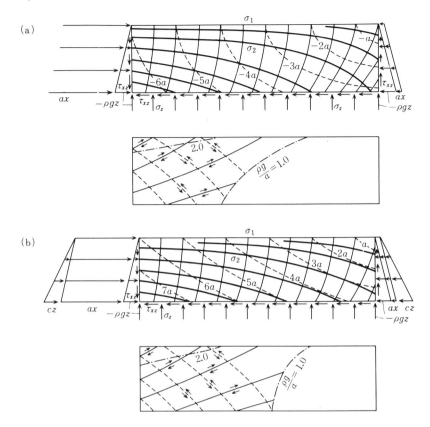

Figure 3.9 Stress trajectories and shear fault pattern developed in a block with horizontal supplementary stress system. (Hafner, 1951* Symbols after Ramsay, 1967). Thick lines in upper diagram of both (a) and (b) are σ_1, thin lines are σ_2, broken lines are maximum shear stress lines. Arrows in lower diagrams show direction of slippage. $-.-.$ shows boundary of stable region corresponding to $\rho g/a = 1$ and 2. (a) Supplementary stress is constant unrelated to depth. (b) Supplementary stress is increasing in proportion to depth

elastic bodies. This model is applicable to a consideration of stress fields in sedimentary cover produced as a result of the vertical movement of the basement. Stress fields around the elastic body are hydrostatic, and because the additional horizontal tectonic stress was not applied, the horizontal stress is simply proportional to the depth. Sanford studied this problem using stress functions and obtained displacement and stress distributions. He analysed three types of displacement, and one of these, namely the case in which

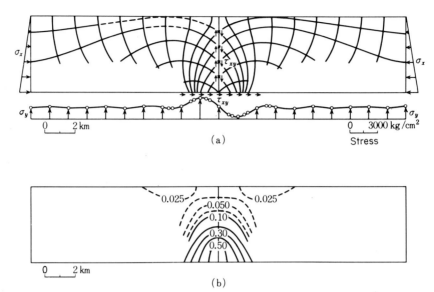

(a)

(b)

Figure 3.10 Stress field generated inside elastic layer resulting from step-like displacement in the basement. (a) Stress distribution (solid line — compression, broken line — tension). (b) Distortional strain energy distribution. Units in kg/cm^2

step-like vertical displacement is applied to the base of overlying strata due to movements of a basement divided by vertical faults, is shown in Figure 3.10(a). The vertical displacement in this diagram is rather exaggerated and in fact would be no more than a mere 10–12 m in an overlying layer 25 km long and 5 km thick. Therefore the direction of the trajectories of the principal stress is mainly determined by the thickness, and the Poisson's number has only a small effect.

Sanford also calculated the torsion strain energy and estimated that fracturing or yielding would probably occur when the values were high (Figure 3.10(b)).

As is evident from Figure 3.10(a) deformation is virtually limited to the area directly above the basement fault. As a field of horizontal tension is created towards the relatively elevated side (left) near the surface, vertical fissuring or gravity faulting develops at this area. Sanford verified these results in experiments using mixtures of clay and sand, following which he found an angle of internal friction of 28° to correspond best with the theory.

The Hafner and Sanford models described above form the fundamental models for any discussion of stress fields arising from two differing causes, namely horizontal compression and vertical uplift. Whether the *Front Range* in Colorado has its structural origins in horizontal compression or vertical

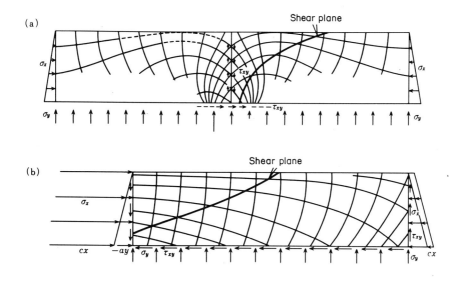

Figure 3.11   Two types of stress fields leading to overthrust faults. (a) Sanford model
(1959*), (b) Hafner model (Harms, 1965*). Curves show trajectories of principal
stress (broken line section, tensile stress)

uplift has been a subject of argument for twenty years. Harms (1965*)
examined the stress field there, using the Hafner and Sanford models, by the
study of the sandstone dyke swarms derived from the Cambrian system,
which intruded the Precambrian system at the time of *Laramide orogeny* (late
Cretaceous – early Tertiary) when the Front Range developed. He found that
whereas there were no places suitable for dyke formation in horizontally
compressed bodies accompanied mainly with upward-concave low angle
thrust systems as in the former case, and that the uplifted bodies with
upward-convex reverse fault systems, as in the latter case, were appropriate
to dyke formation, because they accompanied a stress field of horizontal
tension on the relatively elevated side. Combining this with other facts about
the area Harms concluded that the origins of the Front Range lay in vertical
uplift (Figure 3.11(a) and (b)).

(c)   Superposition of stress fields

Some highly interesting radial dyke swarms are found around the *Spanish
Peak* in Colorado and have been the subject of research by Knopf (1936*) and
Johnson (161*). Odé (1957*) explained that their distribution and formative
mechanisms were attributable to the overlapping of regional and local stress

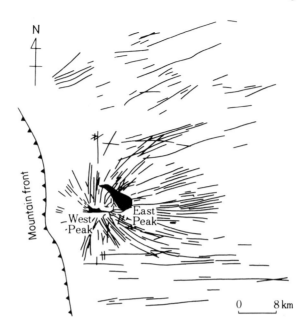

Figure 3.12   Dyke swarms around the Spanish Peaks, Colorado (Knopf, 1936*)

fields. These dyke swarms are distributed almost symmetrically relative to the N70 °E line passing over the West Spanish Peak. There are no fewer than 500 dykes and the majority originate from the West Spanish Peak and tend to turn towards the east. Some are more than 40 km long but those in the west are very short (Figure 3.12). Most of the dykes are vertical and are commonly about 3 m wide. According to the theory of Anderson (1951) the dykes are developed along a plane perpendicular to the minimum principal stress axis, i.e. a plane containing the axes of the maximum and intermediate principal stresses, the problem then becomes to investigate whether the form and position of the principal stress trajectories calculated by applying a suitable model correspond to the actual pattern of the dykes or not.

Because the stress field at the time of the dyke intrusion may be visualized as being composed of local forces brought about by fluid pressure inside a volcanic centre, and regional forces, thought to be uniform across a wide area, the symmetry of the dyke pattern means that the total of these two stress fields possessed similar symmetry. Moreover, because axis-symmetrical radiating dykes can be expected to form from local stress fields the symmetry shown by the actual dykes may be thought of as a reflection of the regional stress field, where the axis of maximum principal stress was perhaps parallel

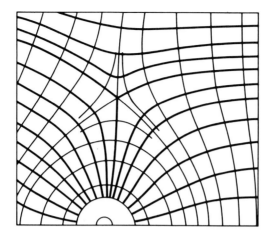

Figure 3.13   Model of stress field around the Spanish Peaks when dyke swarms were formed (Odé, 1957*). Thick lines are trajectories of maximum principal stress. Thin lines are trajectories of minimum principal stress

to the line of symmetry. Moreover the lack of dykes westwards from the mountain front shows that it formed a rigid barrier for both types of stress field. If one simplifies the problem, therefore, into two dimensions, the flat stress field, existing when round holes subjected to hydrostatic pressure opened up near the margins of semi-infinitely large plates, becomes the model for the local stress field.

Finally when one tries to superpose a regional stress field obtained by a stress function on to such a local stress field determined by a similar stress function one obtains the stress trajectories shown in Figure 3.13 which are very reminiscent of the actual dyke pattern.

In each of the three examples introduced above we predicted the fracture pattern which should emerge by determining the position and form of the lines of principal stress. Conversely it enables us to assess the origins of pre-existing fractures.

3.4   The origins and stress fields of regional joints

Joint systems showing uniform directional tendencies across a wide area are known as *regional joints*. Figure 3.14 shows an example of such regional joints with extremely uniform strikes. Such joint systems often tend to be composed of two or three sets, and can be divided into longitudinal joints, i.e. those with strikes parallel to the trend of the major structures of the area; transverse or cross joints, which are at right angles to the major structural trend, and

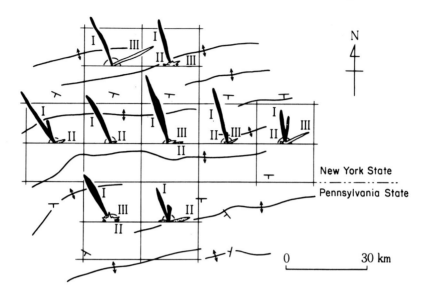

Figure 3.14 Part of regional jointing from southern New York State to northern Pennsylvania in the USA (Muehlberger, 1961*). I, II and III respectively show the strikes of shear joints, tension joints, and joints interpreted as tension joints

oblique or diagonal joints which cross the former sets obliquely. In each case the dips of the joint planes tend towards the vertical. Longitudinal and transverse joints are also both *tension joints*, whereas oblique joints have come to be known as shear joints, although the reason for this is unclear.

Price (1959*) was the first to explain the mechanical origins of regional joints. He assumed that the fracturing of crustal material would be of the brittle type within the limit of elastic deformation and examined whether or not this sort of joint system could be formed solely by the application of compressive stress in a horizontal direction. This is the case when a horizontal compressive stress C_y is applied to Price's standard state of stress described in Section 1. The results were that taking vertical stress due to the overburden as σ_2 and the Poisson's number as m, then C_y is limited to the range:

$$\frac{m-2}{m-1}\sigma_z < C_y < \frac{m-2}{m-1}m\cdot\sigma_z$$

When he examined the fracture criterion from the Poisson's number and coefficient of internal friction obtained in rock tests he found that neither extension nor shear types of fracture could occur in the crust under these stress conditions.

In conclusion, it appears that regional joints are not the products of a period of severe horizontal compressive stress such as a vigorous period of orogenic movements instead, they develop in a subsequent period of weakened horizontal compressive stress by a process in which structures formed during the orogenic movements, rise to shallower levels in the crust. In this case the first thing to be considered is the problem of stress relaxation (see Chapter 6). If the rocks resemble the *Maxwell model*, when the external force is removed, *stress relaxation* will progress until it approaches the hydrostatic state. In the *Voigt model*, however, elastic strain under the applied external force is maintained completely and so a '*residual stress*' showing the precise state of stress at that time will normally be present. On the other hand the progress of upheaval will give rise to tensile stress in the horizontal direction which may be estimated as follows. Taking the radius of the earth as R, the amount of upheaval as z, the horizontal distance of the upheaved area as L and its increase as l, the horizontal tensile stress becomes $l/L \doteq z/R$ from Figure 3.15. Therefore, taking the Young's modulus as E, the tensile stress is given as $T = E.z/R$ from Hooke's law.

The problem of regional joints is thus controlled by the relationship between the vertical stress, which will gradually decrease as the uplift of the structural body progresses, and the newly emergent horizontal stress, which will increase. It is therefore ascribed to tracing changes in the stress field associated with the process of upheaval. As described earlier, there are two cases associated with stress relaxation models dependent on which model the rheological properties of the structural body most closely resemble.

Let us consider the first case in which the stress field at the beginning of the period of lift is close to the hydrostatic state. This corresponds to a case in which the materials comprising the rising structure approximate to the

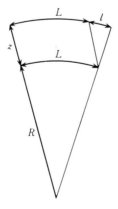

Figure 3.15   Increase in horizontal distance accompanying uplift (Price, 1959*)

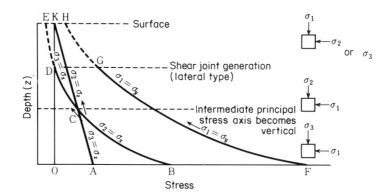

Figure 3.16   Stress history in case when stress field at the beginning of uplift is close to hydrostatic pressure (Price, 1959*). Arrows show directions in which stress changes with the passage of time

Maxwell model. In Figure 3.16 the depth is measured on the vertical axis and stress on the horizontal axis. The point B marks the commencement of uplift. One may also use this diagram with time represented on the vertical axis and the level of point B as time 0 to express the *stress history* of a formation, i.e. changes in stress over time due to the process of advancing upwards uplift. In this case residual stress at point B forms one of the initial conditions. In vertical stress $\sigma_2 = \rho g z$ (ρ average density, g = acceleration due to gravity, z = depth) and is directly proportional to depth, therefore it will decrease with time upwards, along the straight line BA reaching zero (or precisely, atmospheric pressure) at the surface A. The horizontal stress is of three types. The first is produced as a result of overburden and taking the Poisson's number as m it becomes $\sigma_x = \sigma_y = \sigma_z (m - 1)$ as described in Section 3.2(a). This will decrease upwards, along the curve CGA. The second is a residual stress BC of only minimal value. If this is taken to be unrelated to depth but constant throughout the region the curve BD formed by adding BC to the curve CGA will become the total compressive stress in the horizontal direction. On the other hand, the third horizontal stress T is a tensile stress and is proportional to the amount of uplift as previously forecast. It therefore has a negative value and when incorporated into the calculations means that the total horizontal stress will decrease with time along the curve BEF, reaching zero at the point E and subsequently becoming tensile. Because σ_z $>\sigma_x \doteqdot \sigma_y$ as shown in Figure 3.16 the relationship of the axes of principal stress will be $\sigma_1 = \sigma_z$, $\sigma_2 = \sigma_x \doteqdot \sigma_y = \sigma_3$. This is an arrangement of stress axes which will produce the normal faults but when the known physical constants, etc. are examined, it is apparent that no fracturing will occur because the shear stress clearly does not reach the shear strength, even when the differential stres $\sigma_1 - \sigma_3$ reaches its maximum value, FH. Passing the point E the uplifted body reaches area of lateral tension, but as the tensile strength is normally approximately ten times less than the shear strength, then, for example, a tension fracture will be formed at point F. This fracture plane is a vertical plane containing the σ_1 axis, and if σ_2 is exactly equal to σ_3, its strike will be indeterminate. However, the σ_2 and σ_3 axes are in practice distinguished by small amounts of residual stress, and the fracture plane will position itself at right angles to the σ_3 axis. Moreover, as soon as a fracture is produced at point F, tensile stress in the direction σ_3 is released and will develop into compressive stress manifested at point G. Therefore exchange of the σ_2 and σ_3 axes is brought about. If uplift continues a second tension fracture will finally occur at right angles to the new σ_3 axis and the smaller the residual stress the shorter will be the interval of time between it and the first one. Thus the two systems of intersecting tension fractures will each correspond to longitudinal and transverse joints.

Next we consider the second case in which the residual stress is fairly large at the commencement of uplift. This time the rising structural body possesses characteristics similar to the *Voigt model*. As shown in Figure 3.17 the state of stress at the beginning of upheaval is $\sigma_1 = \sigma_y > \sigma_2 = \sigma_x > \sigma_3 = \sigma_z \, \rho g z$, each of which decreases respectively along the curve FG, curve BCD and straight line ACK. σ_2 and σ_3 intersect at point C, after which $\sigma_1 = \sigma_y > \sigma_2 = \sigma_z = \rho g z > \sigma_3 = \sigma_x$. Because the σ_2 axis becomes vertical beyond the point C this provides one of the conditions for the formation of a vertical shear plane of

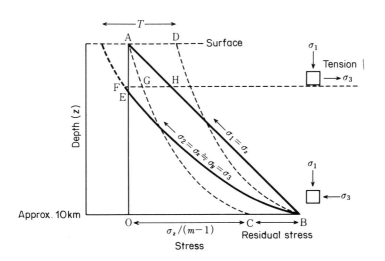

Figure 3.17   Stress history in case when residual stress at the beginning of uplift is fairly large (Price, 1959*). Vertical distance about one quarter of that in Figure 3.16. Arrows show directions in which stress changes with the passage of time

the strike-slip fault type. When the level DG is reached another condition is satisfied (i.e. the restriction for the minimum value of σ_1/σ_3 demanded from the coefficient of internal friction and the cohesion value), and a set of vertical shear planes develop in a direction of $45° - \psi/2$ to the σ_1 axis (ψ is the angle of internal friction). Because most of the residual stress is lost at the same time, subsequently $\sigma_1 = \sigma_2 = \rho gz > \sigma_2 = \sigma_x \doteqdot \sigma_y = \sigma_3$. Exactly the same *stress history* as found for case 1 is followed, and two sets of vertical tension fracture planes are formed successively at right angles to both the σ_1 and the σ_3 axes. Combined with the shear planes in case 2, three types of joints, i.e. oblique, longitudinal and transverse, will systematically occur.

When we look at Price's work, therefore, from the point of view of stress history, we can follow a stress field predominant in an early period through subsequent periods of its being a residual stress and see it play an important role in the generation of new faults as it declines.

Takeshi Uemura and Akira Iwamatsu

Japanese references

Kakimi (1968): On some problems concerning minor faults. *Chinetsu*, **17**. 5–13.
Collaborative Research Group for Niitsu Anticline (1977): On the folding mechanism of the Niitsu anticline, oil field of Niigata, *Japan. Earth Sci. (Chikyu Kagaku)*, **31**, 70–82.
Uemura, T. (1976): Some problems on earthquake and the formation of geologic structures. *Mem. Geol. Soc. Japan*, **12**, 43–9.

References

Anderson, E. M. (1951): *Dynamics of Faulting and Dyke Formation*, 2nd ed., 206 pp., Oliver & Boyd, Edinburgh.
Donath, F. A. (1961): Experimental study of shear failure in anisotropic rocks, *Geol. Soc. Am. Bull.*, **72**, 985–90.
Griggs, D. T. and Handin, J. (1960): Observations on fracture and a hypothesis of earthquake, *Geol. Soc. Am. Mem.*, **79**, 347–64.
Gzovsky, M. V. (1954): *Tektonicheskie polya napryazheniy*, Izd. A. N. SSSR ser. geof., **5**, 390–410.
Hafner, W. (1951): Stress distributions and faulting, *Geol. Soc. Am. Bull.*, **62**, 373–98.
Harms, J. C. (1965): Sandstone dykes in relation to Laramide faults and stress distribution in the southern Front Range, Colorado, *Geol. Soc. Am. Bull.*, **76**, 981–1002.
Hubbert, M. K. (1951): Mechanical basis for certain familiar geologic structures, *Geol. Soc. Am. Bull.*, **62**, 355–72.
Huzita, K. (1969): Tectonic development of southwest Japan in the Quaternary period, *J. Geosci. Osaka City Univ.*, **12**(Art. 5), 53–70.
Jaeger, J. C. (1960): Shear failure of anisotropic rocks, *Geol. Mag.*, **97**, 65–72.

Johnson, R. B. (1961): Patterns and origin of radial dyke swarms associated with West Spanish Peak and Dyke Mountains, south-central Colorado, *Geol. Soc. Am. Bull.*, **72**, 579–90.

Knopf, A. (1936): Igneous geology of the Spanish Peaks region, Colorado, *Geol. Soc. Am. Bull.*, **47**, 1727–84.

Muehlberger, W. R. (1961): Conjugate joint sets of small dihedral angle, *J. Geol.*, **69**, 211–19.

Odé, H. (1957): Mechanical analysis of the dyke pattern of the Spanish Peaks area, Colorado, *Geol. Soc. Am. Bull.*, **68**, 567–76.

Price, N. J. (1959): Mechanics of jointing in rocks, *Geol. Mag.*, **96**, 149–67.

Sanford, A. R. (1959): Analytical and experimental study of simple geologic structures, *Geol. Soc. Am. Bull.*, **70**, 19–52.

Sugimura, A. and Matsuda, T. (1965): Atera fault and its displacement vectors, *Geol. Soc. Am. Bull.*, **76**, 509–22.

Uemura, T. and Shimohata, I. (1972): Neutral surface of a fold and its bearing on folding, *Rept. 24th IGC*, Sect. 3, 599–604.

Further Reading

The following specialist treatments of fracture and failure mechanisms in rocks are available.

Jaeger, J. C. (1969): *Elasticity, Fracture and Flow*, 3rd ed., 268 pp., Methuen, London.

Jaeger, J. C and Cook, N.G.W. (1969): *Fundamentals of Rock Mechanics*, 515 pp., Methuen, London.

Nadai, A. (1950): *Theory of Flow and Fracture of Solid*, 2nd ed. 572 pp., McGraw-Hill, New York.

Nagumo, S.: Rock failure. Earth Sciences Series (Kyoritsu Shuppan Tokyo) Vol. 6, 273–347.

Price, N.J. (1959): *Fault and Joint Development in Brittle and Semi-brittle Rock*, 176 pp., Pergamon, Oxford.

Ramsay, J. G. (1967): *Folding and Fracturing of Rocks*, 568 pp., McGraw-Hill, New York.

Yamaguchi, U. and Nishimatsu, Y. (1977): *Introduction to Rock Mechanics*. 2nd ed. 250 pp. Univ. Tokyo Press, Tokyo.

The problems of fracturing are touched on in most textbooks on structural geology.

Geological Structures
Edited by T. Uemura and S. Mizutani
©1984 John Wiley & Sons Ltd.

4

Rock Cleavage

As one journeys into the southern parts of the Kitakami Mountains, of northeast Japan, one sometimes comes across traditional Japanese houses hung with natural slate (Figure 4.1). These jet black scale-like roofs and walls are very beautiful. Most of them are made of the Permian *Toyoma slate*. The Triassic Inai slate (calcareous laminate slate) as used in inkstones or for paving is also well known as a local speciality.

Geological research was begun long ago in the southern Kitakami Mountains because of its abundant fossils, and it is one of the best surveyed areas in Japan. Ren Kurimoto, who was a technical expert at the Ministry of Agriculture and Commerce and one of Japan's pioneer geologists, made a study of Okatsuhama slate (*Inai slate* from Okatsu in Miyagi Prefecture), and pointed out its tectonic significance in the light of the then new information he had learned at the Royal School of Mines in London (Kurimoto, 1886). Despite this, however, his early work was not followed up, and it failed to attract much academic interest until very recently. This was probably because the mainstream of structural geology at the time lay in the explanation of geological history on the basis of bio-stratigraphical sequences.

It is important to establish the deformation mechanisms and physical and mechanical conditions prevailing during deformation of rocks and strata. *Rock cleavage* is a valuable key to clarify such matters. Thus, in this chapter we outline the properties and origin of rock cleavage from such a point of view. Cleaved rocks are also locally distributed in the Shimanto, Setogawa, Maizuru and Yamaguchi zones in southwest Japan.

4.1 Types of cleavage

Rock cleavage is defined as 'fine planar secondary structures produced by deformation'. Generally speaking cleavage cuts obliquely across the bedding plane and is characterized by a uniform relationship to the axial planes of folds. Particular instances when it is parallel to the axial plane of a fold are known as '*axial-planar cleavage*'. Because schistosity seen in crystalline schists is defined as 'planar structures due to dimensional preferred orientation† of

† Statisically uniform orientation of inequant grains or crystal aggregates.

ERRATUM

We would like to point out that Figure 4.1 on page 162 has been inverted. We apologise for this error.

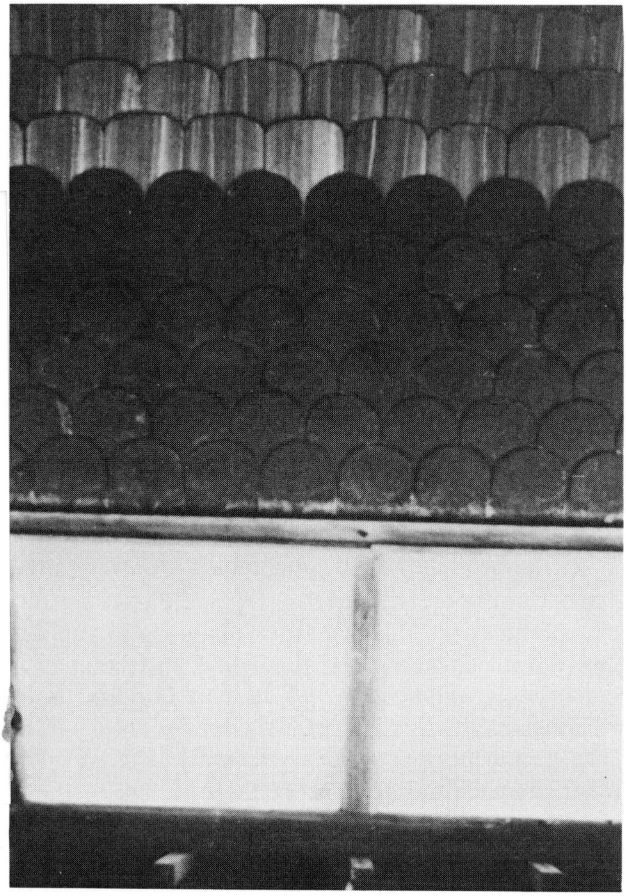

Figure 4.1 Slate on the wall of a country house (Kitakami Mountains). The upper black material is Toyoma slate and the lower banded material is Inai slate (Photo T. Uemura)

foliated minerals', it is sometimes confused with cleavage. However, schistosity has a different mechanical significance as it is often parallel to the bedding plane and is not always related to the axial plane of folds. In this chapter we therefore omit schistosity from the discussion.

The first person to identify and describe cleavage was Bakewell in 1815. In the following 150 years various forms of terminology have been proposed and many types of classification attempted. In this chapter we will try to use only standard terminology and to introduce the minimum essential classifications.

(a) Fracture cleavage

Fracture cleavage refers to very fine fractures or micro-faults at intervals of only a few millimetres or centimetres and is unrelated to the arrangement of the constituent minerals of the rock. It may occur in almost all types of rock but in general it is most commonly seen in deformed non-metamorphosed mudstones. Irregular anastomosing cleavage may be seen in the Pliocene Nishiyama mudstone in Niigata Prefecture of Central Japan. The cleavage planes are at intervals of 2–3 mm and they intersect and converge with each other (Figure 4.2). The orientation as a whole is regular and almost at right angles to the bedding plane, but it is sometimes dragged along due to bedding plane slip. Microscopic examination reveals the cleavages to be filled with a clay film 20–30μ thick and clastic mica flakes can be seen parallel to the cleavage plane in the film. The intersection of fracture cleavage and the bedding plane often runs parallel to the fold axis. Moreover, fan arrangements converging on the centre of curvature are normally seen around the fold axis in a cross-section at right angles to the axis (see the example of sandstone in Figure 4.3). Some geologists explain the origins of fracture

Figure 4.2 Fracture cleavage in the Pliocene Nishiyama formation in the Niitsu oilfield, Niigata Prefecture (Photo T. Uemura)

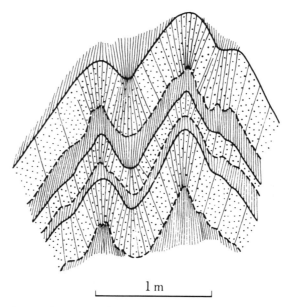

1 m

Figure 4.3   Fanning and 'refraction' of axial plane cleavages (after Hobbs *et al.*, 1976*)

cleavage in exactly the same way as jointing, but Price and Hancock (1972) concluded that it was formed as a result of hydraulic fracturing when the *axis of minimum principal stress* was temporarily parallel to the bedding plane and at right angles to the fold axis, and the *pore water pressure* became abnormally high during folding.

(b)   Slaty cleavage

Slaty cleavage is encountered in fine-grained weakly metamorphosed rocks and is composed of parallel arrangements of very fine, foliated minerals such as illite. Because slates and tuffs showing slaty cleavage can be separated into extremely thin flakes along the cleavage planes, they can usually be cut into thin sheets simply by inserting a chisel. This property means that rocks showing slaty cleavage may readily be made into roofing slates and tiles. It also means that weathering tends to emphasize the cleavage plane rather than the bedding plane and so particular care must be taken to distinguish between the two in the field.

When slaty cleavage is examined under the microscope various lenticular areas, $10–50\mu$ thick, are found to be arranged throughout with a remarkable

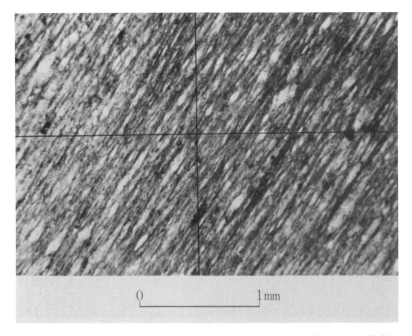

0   1 mm

Figure 4.4   Photomicrograph of slaty cleavage in the upper Devonian Tobigamori
formation (Kitakami Mountains)

degree of preferred orientation and extremely thin films are seen surrounding
them (Figure 4.4). As these film sections are composed of foliated minerals
arranged almost parallel to the direction of cleavage they show parallel
extinction under crossed Nicol prisms. Clay minerals, ilmenite or zircon
sometimes accumulate in the film so it can easily be distinguished as black
stripes in plane polarized light. The film is especially well developed around
large crystals and rock fragments enclosing them and consequently may often
form a pressure shadow† (Figure 4.5). The lenticular areas may be composed
of aggegates such as quartz, feldspar or rock fragments and these may
occasionally show rotation. Foliated minerals are usually absent but if present
they show no particular orientation.

In its relationship with folding, slaty cleavage is commonly said to be
parallel to the axial plane of so-called shear folds and morphologically *similar
folds* are formed by slight slippage along the plane of cleavage (Figure 4.6). A
feature of shear folding is thus that the thickness of the bed measured parallel
to the axial plane of a fold, i.e. the *axial plane thickness*, is constant. When

†Texture indicated by later growths of fine mineral aggregates on opposite sides of a host
porphyroblast in crystalline schists or detrital grains in slates.

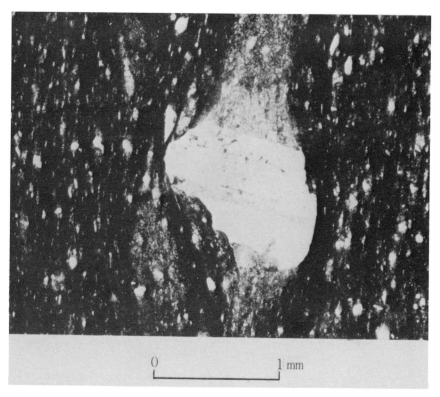

Figure 4.5 Photomicrograph of pressure shadow in the Permian Toyoma slate (Kitakami Mountains)

running through competent layers such as sandstones, the slaty cleavage is refracted at the boundary between these and mudstone and a cross-section at right angles to the fold axis normally reveals convergent fans, near the fold axes, facing the centre of curvature in the sandstone, and conversely, divergent fans, near the fold axes, facing the centre of curvature in the sandstone, and conversely, divergent fans in the mudstone (Figure 4.3).

(c) Crenulation cleavage

Crenulation cleavage is seen in low- to medium-grade metamorphic rocks with well developed bedding plane schistosity. The bedding plane schistosity shows extremely fine folds with wavelengths of a few hundred microns with mutually parallel axial planes, which repeat themselves over and over and the crenulation cleavage runs parallel to these axial planes. Although a number of different forms of crenulation cleavage exist (Hara, 1966; Iwamatsu, 1971,

Figure 4.6 Minor folds in Inai slate (Ojika Peninsula, Kitakami Mountains)

1975*) they are gradual developments from each other and are very closely related in origin. Broadly speaking one may distinguish two basic types or end members. One is *discrete crenulation cleavage* which clearly cuts and displaces the schistosity; and the other is *zonal crenulation cleavage* in which the cleavage is zone with diffuse boundaries which corresponds to the parallel limbs of the microfold with a wavelength measured in microns, and particularly to shorter limbs that are steeply inclined (Gray, 1977) (Figure 4.7). In this section we discuss the former case which is the more common type of crenulation cleavage (Figure 4.8, Figure 4.7 (b)). This type of crenulation cleavage is more irregular and intermittent than slaty cleavage and may also become branched. Microscopic examination reveals minute sericite flakes arranged parallel to each other and running parallel or at a slight angle to the cleavage. These may sometimes become curved as a result of dragging along the cleavage particularly when they run obliquely. They may sometimes even be connected with the schistosity. In other words it turns into zonal crenulation cleavage. The bedding plane schistosity is shifted little by little as a result of such cleavage, forming minor folds with wavelengths of a few millimetres throughout the material. Therefore the sense of slippage along the cleavage is reversed on the two limbs of these minor folds (Figure 4.7(c)).

There are many examples of areas in which the folds from the smallest to the longest wavelength are effectively composed of aggregations of such minor folds, and in the majority of cases such variously sized folds all have mutually parallel axial planes. These fold axial planes thus run parallel to the crenulation cleavage. In this sense there is a suggestion that it plays the same role as slaty cleavage and that the two may be closely connected in origin.

We have briefly introduced three types of cleavage above and in the following section we concentrate on slaty cleavage which is of the greatest interest in terms of deformation theory, because slate has many *strain indicators* such as deformed fossils, etc.

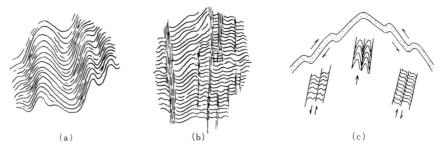

(a)   (b)   (c)

Figure 4.7   Types of crenulation cleavage (Rickard, 1961). (a) Zonal crenulation cleavage. (b) Discrete crenulation cleavage. (c) Opposite senses of rotation indicated by micro drag-folds and cleavage plane slip

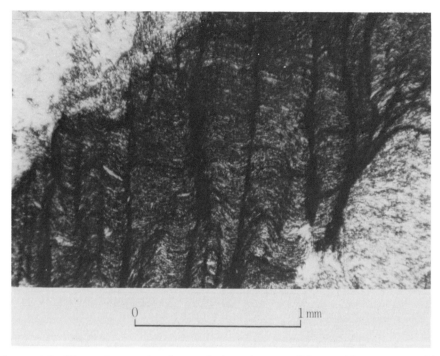

Figure 4.8   Photomicrograph of crenulation cleavage (Matsugahira metamorphic rock, Abukuma Mountains)

4.2   The origins of slaty cleavage

Since the beginning of the nineteenth century, research into slaty cleavage has been increasing. Geometric and morphological surveys of slaty cleavage appeared in the last century and there were repeated arguments about its origins in the light of such work (Leith, 1905). The arguments continued later on but without bearing any real fruit or reaching any firm conclusions. This was mainly due to the then rather unsophisticated levels of knowledge in the science of rock deformation and metamorphic petrology, but also lay in the rather sweeping use of the term 'slaty cleavage', giving no consideration to the variety or degrees of deformation or metamorphism in the rocks. For example, there was an argument over whether foliated minerals found parallel to slaty cleavage were originally clastic materials which had rotated (Sorby, 1853) or whether they were new minerals produced by recrystallization when cleavage developed (van Hise, 1896). Currently a compromise between the two ideas seems dominant. For example, Ramsay (1967*) and Williams (1977*) explained that mechanical rotation is the principal factor at work in conditions of very low metamorphism, but that the importance of recrystallization increases with temperature. Ono (1973*) also studied slaty cleavage in the Palaeogene Setogawa group and concluded that it is a product of both mechanical rotation and recrystallization. Oertel (1970) gave a theoretical explanation in which a foliated mineral acquires a mechanically rearranged orientation and this is then emphasized further by crystal growth.

There are currently three major theories concerning the origins of slaty cleavage, i.e. the shearing hypothesis, the flattening hypothesis and the tectonic dewatering hypothesis. In most respects the flattening hypothesis probably has the most advocates.

(a)   Shearing hypothesis

The *shearing hypothesis* was first propounded by Phillips (1844) and Becker *et al.* (1896, 1904, 1907). Becker considered it in terms of strain ellipsoid theory and said that the plane of maximum shear stress is where the cross-section of strain *ellipsoid* forms a perfect circle and that slaty cleavage will develop parallel to this plane. Tokuyama (1971*) studied the internal structure of the Toyoma slate and discovered bi-directional cleavage planes intersecting each other at an angle of 15°. He also found that the direction of the bisector of the obtuse angle between the two contained the principal axis of maximum compressive stress (Figure 4.9). This led him to emphasize that slaty cleavage initially forms as *conjugate shear planes* and, as the strain develops, rotates

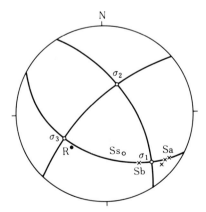

Figure 4.9   Stereographic projection of Toyoma slate in the Kitakami Mountains (Tokuyama, 1971*). Sa, Sb: poles of slaty cleavage planes. Sa $\wedge$ Sb = 19°. Ss: poles of bedding plane. R: tensile crack. σ_1 σ_2 σ_3: principal axes of maximum, intermediate and minimum compressive stress

until the currently apparent obtuse angle is achieved. However, slaty cleavage normally develops in a single direction and even if the arrangement of the minerals had been bi-directional there must be doubt as to whether this would be indicative of a shearing action. If a shear fracture develops along a particular plane it is thought that orientation of subsequent fractures will be controlled by the early anisotropic planes. It would be very unlikely that shear fracture planes would form several tens of μ apart as happens in slaty cleavage.

(b)   Flattening hypothesis

The flattening hypothesis has had a great many advocates, from Sharpe (1847, 1849) and Sorby (1853, 1856) to Wood (1973, 1974*), Tullis (1975, 1976) and Siddans (1972*, 1977) today. Sharpe (1847) studied deformed fossils in slate and concluded that the cleavage is at right angles to the direction of maximum pressure, as the fossils are flattened into the plane of the slaty cleavage. *Strain analysis* discussed later) was subsequently performed by many researchers in this field, using various strain indicators such as deformed fossils or pebbles and it was emphasized that slaty cleavage develops in a plane normal to the principal axis of maximum compressive strain. Dieterich (1969*) used the finite element method for a computer simulation of buckling folds in viscous sheets in viscous media (see Chapter 5). He thus demonstrated that the arrangement of planes intersecting the maximum compressive strain corresponds remarkably well with the way in which slaty cleavage is arranged in nature. (Compare Figures 4.10 and 4.3.)

(c)   Tectonic dewatering hypothesis

Maxwell (1962*) studied the *Martinsburg slate* in the Appalachian Ordovician system, and noted that clastic dykes in the slate run parallel to the slaty cleavage (Figure 4.11), he concluded that the cleavage developed as a result of the mechanical rotation of clastic particles in the dewatering process, when the pore water pressure was at an abnormally high level, due to tectonic stresses before the rock consolidated. This is the tectonic dewatering hypothesis. Supporters of this hypothesis of cleavage development during diagenesis which was quite at odds with the then conventional wisdom gradually appeared on the basis of subsequent field studies (Powell, 1972; Altermann, 1973, etc). Counter-arguments were also produced to the interpretation of the facts on which this idea was based (Geiser, 1975). Many facts are revealed in field studies which cannot be explained adequately by this hypothesis. For example, numerous dykes and lavas are found in the Mesozoic and Palaeozoic strata in the Kitakami Mountains. Slaty cleavage has developed in these in the same direction as cleavage in the surrounding slate (Iwamatsu,

1969*, 1975*). In places there is also slaty cleavage in the fossils and the pebbles contained in the slate. It is clear that these were consolidated when the slaty cleavage formed. Superimposition of slaty cleavage in two directions formed at differing periods has been observed in many places throughout the world such as in Taiwan (Kimura, 1973). In such cases the rock had obviously already consolidated when at least the later slaty cleavage developed. Hirowatari and Katayama (1973) used electron microprobe analysis (EPMA) to study the chemical and mineral composition of the Palaeogene *Nachiguro mudstone* and *Toyoma slate*. They established that whereas the Nachiguro mudstone, in which cleavage is never seen, is hardly metamorphosed even though affected by diagenesis, recrystallization effects are fairly well advanced in Toyoma slate. In other words, they showed that slaty cleavage is not formed solely in diagenesis.

Thus the tectonic dewatering hypothesis is difficult to accept as a mechanism for regional slaty cleavage development, but cleavage believed attributable to dewatering may be found at the very local level (Williams, 1977* *et al.*) Moore and Geigle (1974) found slaty cleavage in cores drilled from the deep ocean floor and this could well have been formed by a dewatering process.

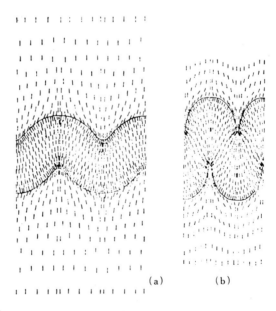

(a)   (b)

Figure 4.10 Computer simulation of buckling folds of a viscous plate in a viscous media (Dieterich, 1969*). Short lines are drawn perpendicular to maximum compressive strain. (a) 100% average compressive strain; (b) 150% average compressive strain

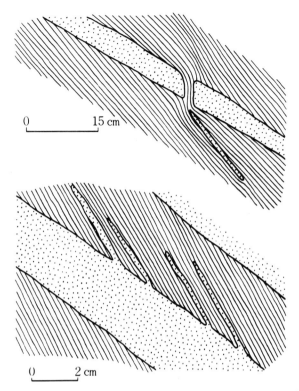

0   15 cm

0   2 cm

Figure 4.11   Sandstone dykes parallel to slaty cleavage in the Martinsburg slate
(Maxwell, 1962*)

4.3   Strain analysis and slaty cleavage

Unlike highly metamorphosed rocks such as crystalline schists, physical
bodies contained in the strata since the time of deposition such as fossils or
ooids† are well preserved in slate. If the original form of these objects is
known they may be used as *strain indicators*. However, unless the viscosity
ratio between the objects and the matrix is known it is not possible to
establish the degree of strain to which the rocks themselves have been
subjected. As the viscosity ratio at the time of deformation is extremely
difficult to establish, it is preferable to focus on objects with properties similar
to the matrix, such as fossil imprints or ooids in limestone. Various strain
indicators are available as long as such conditions are borne in mind. Taking
the Kitakami Mountains as our example, one may use *reducation spots*‡

†Small (0.5 – 1 mm diameter) spherical accretionary body resembling fish eggs in calcareous
  rocks. Deposited as concentric spheres around a nucleus in shallow water.
‡Light green marks in purple-red tuff — said to be formed by iron reduction during diagenesis.
  Completely spherical at the time of formation. Commonly a few millimetres to one centimetre
  in diameter.

Figure 4.12   Reduction spot (20 mm long axis) seen in Devonian Tobigamori red-purple tuffs in Kitakami Mountains. Showing profile almost parallel to X–Z plane. Line of long axis is cleavage plane. Line extending upper left to lower right is bedding plane

(Figure 4.12) in Devonian purple-red tuff, spirifers in Devonian-Carboniferous systems, Permian Usuginu-type conglomerates (Iwamatsu, Yoshihara, 1974; Ikeda, 1977) (Figure 4.13), or Triassic, ammonites (Hayami, 1961).

(a)   Direction of strain ellipsoids and slaty cleavage

Various methods of *strain analysis* have been proposed based on the configurations and properties of strain indicators (Ramsay, 1967*; Ikeda, Shimamoto, 1975). Methods have also been devised which assess the amount of strain from the degree of preferred orientation of the foliated minerals which have formed the slaty cleavage (March, 1932; Tulles and Wood, 1975; Willis, 1977). Whichever method of analysis is employed, the amount of strain is normally given ultimately as a *strain ellipsoid* (Figure 4.14). Assuming that a physical object, which was spherical prior to deformation, is subjected to uniform deformation, one examines the shape of the ellipsoid into which it changes. Taking the lengths of the principal axes of the strain ellipsoid as being in the order of $X > Y > Z$, the plane of maximum flattening (X–Y plane) without exception corresponds to the plane of slaty cleavage in all the orogenic belts in the world. This is the basis for the flattening hypothesis discussed in Section 4.2. Allochthonous fossils, which are normally deposited parallel to the original bedding plane, are also found to be arranged parallel to the direction of slaty cleavage in the Kitakami Mountains. The pebbles of *Usuginu-type conglomerates* similarly have planes of maximum flattening which correspond to the slaty cleavage. This is because they were deformed during folding and were rotationally rearranged. They do not show an imbrication formed during deposition, and therefore, if the strata are returned to the horizontal, in order to establish the paleo-current and to reconstruct the sedimentary basin, one is bound to make the mistake of imagining that the land supplying the material was located in the position of the present anticlinal axis. When reconstructing the paleo-environment in

Figure 4.13   Deformed Usuginu type conglomerate in Toyoma group (Kitakami Mountains)

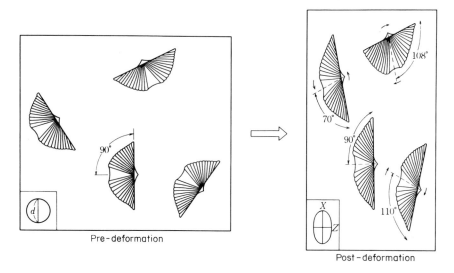

Figure 4.14   Deformation of Spirifer and the strain ellipse (prepared from Ragan, 1973). The feature of pre-deformation Spirifer is the angle of 90° between the axis of symmetry and hinge line

areas with slaty cleavage, very careful consideration must be given to subsequent deformation.

The direction of the long (X) axis differs from one orogenic zone to another, and although it is almost horizontal and parallel to the fold axis in the Kitakami Mountains (Iwamatsu, 1975*), it is almost vertical and at right angles to the fold axis in the Cambrian slate belt in Wales (Ramsay, 1962*) or the Blue Ridge zone in the Appalachians (Cloos, 1947). We do not know the reason for such differences.

(b)   Deformation plots

Let us now consider what kind of information about deformation can be deduced from the lengths of the axes in strain ellipsoids. Various graphs have been suggested to express this two-dimensionally (Flinn, 1956; Hossack, 1968), these being called 'deformation plots' (Flinn, 1956, 1962). In this section we introduce the Wood (1974*) version which is a refinement of that of Flinn (1956). However, care is needed to avoid confusion as the definitions of Flinn treat the strain axes in quite the opposite manner, i.e. $X<Y<Z$. In Figure 4.15 the normal logarithm for the axial length ratio X/Y between the axes for maximum and intermediate principal strain is on the ordinate, and the axial length ratio Z/Y between the axes for minimum and intermediate

principal strain is on the abscissa. Therefore, when strains in the direction of the principal strain axes are equal, a straight line is produced on the graph. The figures on the straight lines show by how many per cent relative to the diameter d of a sphere of equal volume to the strain ellipsoid, each principal strain has increased or decreased. For example, $+ 150$ per cent X indicates a 150 per cent elongation in the direction of the X axis, i.e. $X = 2.50\ d$ whereas -70 per cent Z means a 70 per cent contraction in the direction of the Z axis, i.e. $Z = 0.30\ d$. If there is a reduction in volume accompanying the deformation, a comparison cannot be made with d, but this may probably be ignored in deformation associated with slaty cleavage (Wood, 1974*). Moreover, the line at 45° which passes through the origin expresses plane strain when $Y = d$, and so the area to the lower right of this line indicates disc-shaped deformation of the flattening type (oblate), and to the upper left, rod-shaped deformation of the constriction type (prolate).

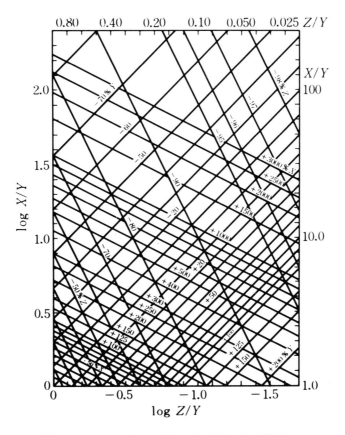

Figure 4.15   Deformation plot (Wood, 1974*)

According to Wood's (1974) strain analysis using *reduction spots* the average strain ellipsoid in the Cambrian slate belts in the Caledonian Orogenic belt is $X : Y : Z = 1.7 : 1 : 0.26$, i.e. $X = 2.23\ d$, $Y = 1.31\ d$ and $Z = 0.34\ d$. In contrast to this, the Taconic slate belt in the Appalachian orogenic belt is $X : Y : Z = 1.7 : 1 : 0.17$, i.e. $X = 2.57\ d$, $Y = 1.51d$ and $Z = 0.26\ d$. The difference probably lies in the fact that the former comprises open folds with almost vertical axial planes whereas the latter consists of closed overfolds. Although the types of folding affect the configuration of strain ellipsoids, the values for the two cases should in fact be said to correspond extremely well. Figure 4.16 was obtained when more than 5000 measurements taken for the Cambrian and Ordovician systems in both areas were all plotted. There is clearly a distinct area of overlap, illustrating the existence of an area over which slaty cleavage is formed (the area enclosed by the dotted line). The values obtained by the author for Devonian reduction spots in the Kitakami Mountains $X : Y : Z = 1.5 : 0.29$, i.e. $X = 1.98d$, $Y = 1.32d$, and $Z = 0.38d$ also fall within the same area. The fact that these values are closer to those of the Caledonian than the Taconic slate belt is no coincidence, but is attributable to the fact the local folding in the Kitakami Mountains is of the

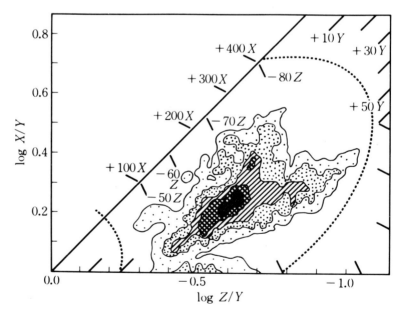

Figure 4.16 Contoured deformation plot of finite strain in slates of Cambro-Ordovician age from Caledonian and Appalachian orogenic belts (Wood, 1974*) Contours 0.5–1–2–3–5% (number of measurements 5200)

open type. The average of all values for both the Caledonian and the Appalachian orogenic belts gives $X : Y : Z = 1.76 : 1 : 0.24$, i.e. $X = 2.35d$, $Y = 1.35d$ and $Z = 0.32d$. Slate with values like these is said to be of the best quality for roofing tile purposes.

In orogenic belts differing in their age and location but showing slaty cleavage we thus find that there is always an approximate 70 per cent contraction in the direction of the Z axis, i.e. the direction at right angles to the cleavage plane. Within the cleavage plane one observes a 100–150 per cent extension in the direction of the X axis and approximately a 35 per cent extension in the direction of the Y axis. What this means is important. Because slaty cleavage throughout the world is steeply inclined and almost vertical, the direction of maximum contraction is almost horizontal (Section 4.4). Therefore, the sedimentary strata in slate belts must have been about three times their present width before deformation, and then must have been subject to orogenic movements based on horizontal compression. At present there is controversy between the horizontalists who believe in generalized horizontal compression and the verticalists who see vertical block movement of the basement as crucial factors in the basic tectonic forces leading to orogenic movements. The above facts provide us with some vital clues for this problem.

Because of the elongation in the direction of the X and Y axes the strata in slate belts must clearly have been thicker than before deformation. Changes in layer thicknesses due to deformation can be up to 1.5 times, when the X axis is nearly vertical as in the Caledonian and the Appalachian orogenic belts. Because it is also common for the amount of horizontal compression to have been greater around the fold axis than on the limbs, combined with the fact that these are morphologically similar folds, there is also apparent thickening of the strata towards the fold axis from the limbs (Figure 4.17).

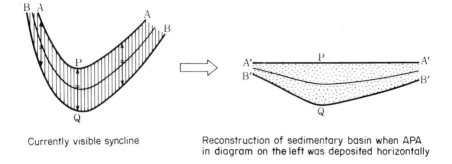

Currently visible syncline

Reconstruction of sedimentary basin when APA in diagram on the left was deposited horizontally

Figure 4.17 Example of conventional reconstruction of a sedimentary basin in the case of shear folding

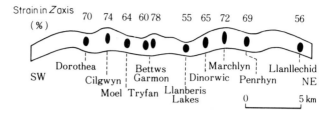

Figure 4.18 Relationship between variation in fold plunge and tectonic thickening as a function of differential strain along the length of the Cambrian slate belt of Wales (Wood, 1974*)

Unless these points are borne in mind there is a risk that the sedimentary basin at the time of deposition may be reconstructed on the basis of the present apparent thickness. It is interesting to observe that in the Cambrian slate belt in the Caledonian orogenic belt, the fold axis forms culminations where the amount of horizontal compression was relatively great, and forms depressions where it was relatively less (Figure 4.18). The reason for this is that when deformation resulted in a thickening of the strata, downwards extension was inhibited and so extension had to take place in an upwards direction (Figure 4.19).

(c) Deformation plots and the formative processes of slaty cleavage

The strain ellipsoids described above show the final results after deformation, but how can we find out by what process the present situation arose from the

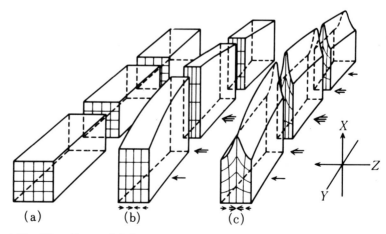

Figure 4.19 The effects of deformation of a rock prism (a) by variable flattening in the X–Y plane (b) and by variable flattening in both X–Y and the X–Z planes

non-deformed sphere? As the origin of a deformation plot represents $X = Y = Z$, i.e. non-deformed sphere, all *deformation paths* should be expected to pass through the origin of the graph. $X = Y > Z$ and $X > Y = Z$ show that the sphere is flattened into a disc shape and elongated to become a rod shape respectively, so the path of the former case will shift from the origin of the graph along the abscissa and the one of the latter along the ordinate. Furthermore, if Y is normally invariable, i.e. $Y = d$, it will progress along a line inclined at 45° through the origin as described above. However, what is the usual form of the path other than in these three special cases?

The values used earlier were averages of values for individual outcrops and some scatter is in fact the normal situation. It is quite natural that one finds a scatter of strain values within individual blocks of only 1000 cm^3 not only in conglomerates containing the pebbles of various rock types, but also in materials thought to be of a fine and uniform structure such as the spots in spotted slate. The greater the degree of deformation the less obvious is the scatter and the greater the harmony between values. In short, when a rock is put under strain by being subjected to a certain force, all the strain indicators are not deformed to the same extent straight away. Some are deformed markedly and others less so as a result of only slight differences in their physical properties, size and concentration relative to the matrix. The scatter of values is produced in this way. However, if the strain progresses, all indicators may be expected to take on similar forms. In other words, the various strain ellipsoids seen today may be said to show the path taken by the most deformed ellipsoid in that block. Figure 4.20 plots the measured

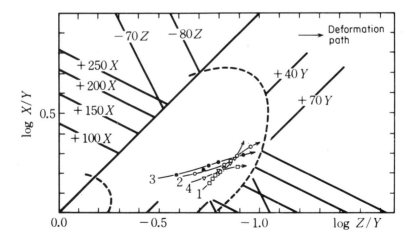

Figure 4.20 Deformation paths for the Taconic slate belt predicted from strain heterogenicity of single localities. 1, 2, 3, 4 indicate locations where samples were gathered (Wood, 1974*)

values at several places in the Taconic slate belt. When points from the same location are joined all rise along smooth curves. These paths also indicate the strain history of the rock. Furthermore the fact that the extensions of these curves seem to pass through the origin confirms the validity of deducing the deformation path in such a way. The accuracy of the method has been further demonstrated by interrupting high pressure deformation experiments using oolitic limestone at various strain levels and examining the deformation of the ooids contained therein.

What do such deformation paths actually mean? The diagram indicates that deformation paths do not follow polygonal lines or complex curves, but trace out pure smooth curves. There is no particular concentration in the distribution of points on the curve, and they are scattered fairly evenly. This shows that the development of slaty cleavage either proceeds steadily over a relatively short time or at least that there are no violent changes of the stress field during the process. This fact is highly significant when considering the progression of orogenic movements.

(d)   The progression of strain and period of slaty cleavage formation

The deformation paths described above only illustrate the route taken after slaty cleavage begins to develop. Reduction spots are said to be formed as a result of a reaction between water and sediments soon after deposition and so it is vital that we know their whole history of deformation from the time of deposition to the development of folding. Beutner (1978*) put forward the following hypothesis with respect to the *Martinsburg slate* (Figure 4.21). The first stage is sedimentary compaction. In the case of mudstone there is commonly a volume reduction of some 50 per cent (Ramsay and Wood, 1973) and the thickness of the layers is halved. Thus the initially spherical object becomes a disc with an axis ratio of $1.0 : 1.0 : 0.5$. If we take the direction that will become the X axis of the future strain ellipsoid, i.e. the vertical, as x, the direction of the Y axis, i.e. of the fold axis as y and the direction of the Z axis, i.e. of horizontal compression as z, the axis ratio of this disc may be expressed as $x : y : z = 0.5 : 1.0 : 1.0$. The second stage is layer parallel shortening which precedes buckling folds. Taking horizontal shortening of 25 per cent as a moderate estimate, the thickness of the layer will be somewhat restored and the previous disc will become one with a ratio of $x : y : z = 0.67 : 1.0 : 0.75$. During this same phase, there will be some tectonic compaction parallel to the layer (Scott *et al.*, 1965) and the density of the rock will rise from 2.5 (shale) to 2.8 (slate). Because this reduces the volume by 10 per cent, the disc will become a prolate spheroid in which $x : y : z = 0.67 : 1.0 : 0.67$. In other words, a section cutting across the fold axis ($X - Z$ plane) will be round. The third stage involves the formation of buckle folds. The limbs of the fold rotate and the layer becomes inclined but because the previous spheroid has a long

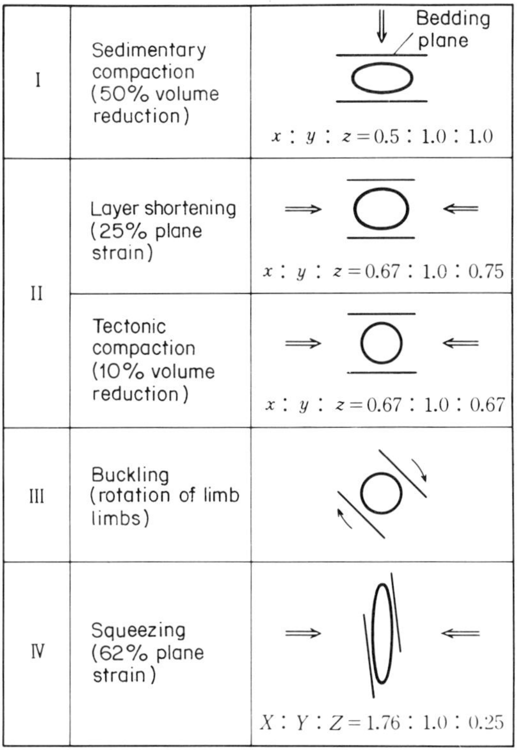

I	Sedimentary compaction (50% volume reduction)	Bedding plane $x : y : z = 0.5 : 1.0 : 1.0$
II	Layer shortening (25% plane strain)	$x : y : z = 0.67 : 1.0 : 0.75$
	Tectonic compaction (10% volume reduction)	$x : y : z = 0.67 : 1.0 : 0.67$
III	Buckling (rotation of limb limbs)	
IV	Squeezing (62% plane strain)	$X : Y : Z = 1.76 : 1.0 : 0.25$

Figure 4.21   Deformation phases leading to production of slaty cleavage; ellipses are sections through the strain ellipsoids perpendicular to the (potential) cleavage plane and the fold axis. See text for explanation (prepared from Beutner, 1978[*])

axis parallel to the fold axis, the axial ratio x, y and z does not change. The fold is of the flexural-slip type accompanied by bedding slip. As rotation of the limbs progresses and reaches the critical point, flattening will occur and slaty cleavage will develop. This is the fourth stage. The amount of flattening imposed in the development of slaty cleavage is 53 per cent in the case of the Martinsburg slate (Beutner, 1978[*]) but is 62 per cent in common roofing slates. Therefore, the previous spheroid becomes an ellipsoid in which $X : Y : Z = 1.76 : 1.0 : 0.25$. This corresponds extremely well with the measurements obtained by Wood (1974[*]) and implies that the hypothesis is probably correct. In other words, so-called shear folds develop initially as flexural slip folds and slaty cleavage develops when the folds are emphasized by flattening. The folds are not formed solely by minute slippage along the cleavage planes.

4.4   Where slaty cleavage develops

Rocks exhibiting slaty cleavage are distributed widely throughout the world although they are in fact limited to special positions in orogenic belts.

(a)   Distribution of foliated structures in orogenic belts

Orogenic belts are usually divided into several structural zones running parallel to the regional trends of the belt. There is some variation due to orogenic type such as island arcs or collision types, but the following kind of division is common from the central axis with its extensive distributions of plutonic rocks to the marginal basins. The names in brackets refer to the corresponding part of the Appalachians which is one of the most typical orogenic belts. Basically the belt is divided into an axial zone (Piedmont zone), internal zone (Blue Ridge zone), external zone (Valley and Ridge zone) and a marginal zone (Allegheny plateau). The axial zone is characterized by bedding plane schistosity which is almost horizontal or gently inclined. There may also be some superimposition of almost vertical crenulation cleavage. The internal zone is characterized by the development of regional cleavage, and the cleavage planes commonly run parallel to the trend of the orogenic belt and are steeply inclined to almost vertical. Because slaty cleavage is most marked in areas with particularly extensive distributions of fine grained rocks, they are called the slate belts. Foliation such as cleavage is not encountered beyond these areas and so the outer margin of the internal zone is known as the tectonite front (Fellows, 1943). Although the boundary between the two zones may in some cases be delineated by major overthrust, the two zones may be structurally related to each other. The lithologies themselves, from the axial to the external zones, tend to be arranged in the order gneisses → crystalline schists → phyllite → slate → shale.

(b)   Depth at the time of slaty cleavage development

As outlined above, the distribution of foliated structures varies according to their position within the orogenic belt. Is this simply because different forms of deformation occur at horizontally distinct locations? Or conversely, does this distribution possibly illustrate some sort of mechanical inevitability?

Generally speaking, factors governing the form of deformation in rocks may be thought to be the physical properties of the rock itself, together with temperature, pressure (confining and tectonic) and strain rate. Taking a single folded structure one may regard the tectonic pressure and the *strain rate* as being almost equal at least for deformation occurring in the same period. In this case, if the fold is composed of strata of the same lithology, deformation seen therein may be taken as a function of the temperature and the confining pressure, i.e. the depth. Below we discuss an example of this line of research (Iwamatsu, 1969*, 1975*).

The Tsunakizaka syncline in eastern Kensennuma in the southern Kitakami Mountains is composed of Permian to Cretaceous strata. Excluding the Cretaceous system, mudstones developed in almost all the horizons in accordance with the conditions outlined previously. These strata formed a huge syncline with the same fold axis. Typical shear folding with axial planar slaty cleavage may be seen in the Permian and Triassic systems, flexural folding accompanied by bedding plane slip in the Upper Jurassic system, and forms intermediate between the two in the Middle Jurassic system (Figure 4.22). In short, from deeper to shallower horizons shear folding gradually

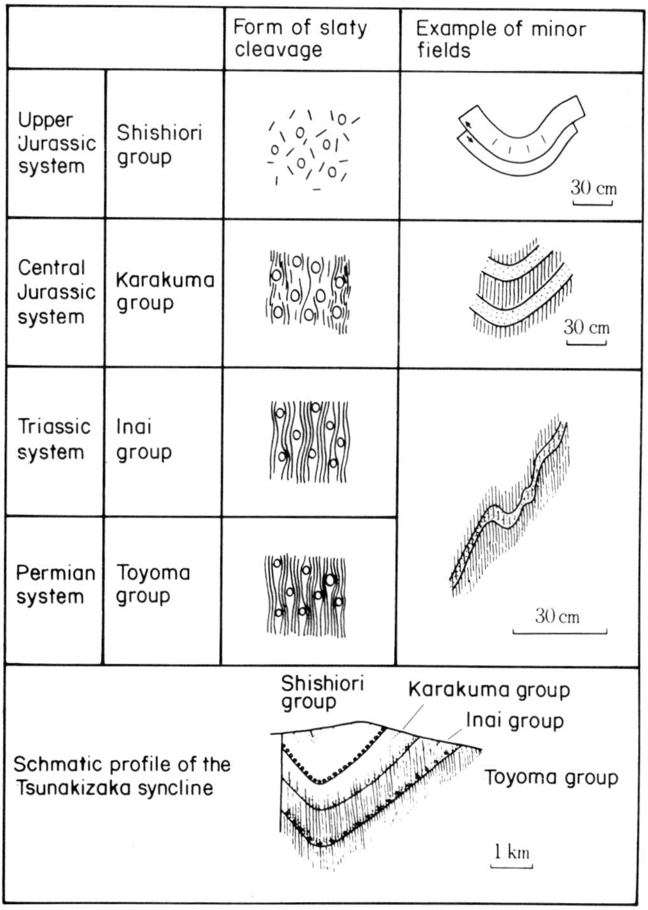

		Form of slaty cleavage	Example of minor fields
Upper Jurassic system	Shishiori group		30 cm
Central Jurassic system	Karakuma group		30 cm
Triassic system	Inai group		
Permian system	Toyoma group		30 cm
Schmatic profile of the Tsunakizaka syncline		Shishiori group Karakuma group Inai group Toyoma group	1 km

Figure 4.22 Minor structures of Tsunakizaka syncline in Kitakami Mountains (Iwamatsu, 1975*)

changes to flexural folding. The frequency of the slaty cleavage per unit
length, measured in the plane at right angles to the cleavage, is also seen to
decrease from the deeper to the shallower horizons (Figure 4.23). Its
properties are remarkably regular and uniform in the Permian and Triassic
systems but in the Middle Jurassic system they become irregular and
discontinuous (there are marked differences in the development of cleavage
even within the same thin section), whereas cleavage is frequently not found
at all in the Upper Jurassic system. Some parallel unconformity or slight
angular unconformity is found below the Triassic and the Jurassic systems,
but there is no sudden change in the structures at these boundaries and no
superimposed deformation on the strata below. There are, therefore, no
particular grounds for believing that major crustal movements such as the
Honshu orogenic movement inferred by some authors took place in this
period. Consequently all the above-mentioned structures are thought to have
developed during Cretaceous orogenic movement. Thus, even when subjected to identical tectonic movements, different forms of deformation will be
produced depending upon depth. At depth, shear folding with axial planar
slaty cleavage will form, whereas in shallower horizons flexural folding will

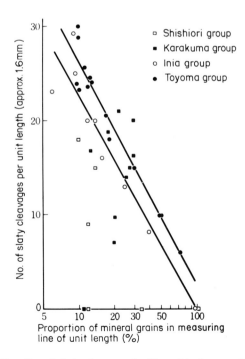

Figure 4.23 Density of slaty cleavage in Tsunakizaka syncline (Iwamatsu)

develop. Let us consider the location of slaty cleavage development in view of the above conclusions. It is thought that at the time of the Cretaceous orogenic movement, the Permian and the Triassic systems were deeper even than the current thickness of the Jurassic and the Cretaceous systems, of about 3000 m. Slaty cleavage is thus thought to have developed at depths greater than this and under confining pressures of 700–800 bar. Typical slaty cleavage also formed in the Permain and the Triassic systems in the Ojika Peninsula in the Kitakami Mountains and the 'depth' here was put at about 2500 m (Tokuyama, 1965*). Similar values of about 3000 m were reported for samples from widely differing locations and periods. According to Carson (1968) load pressures for depths up to 3000 m were of significance in the development of cleavage in the *Martinsburg slate*. Incipient slaty cleavage was also discovered in Miocene shales, 3500 m deep, in the core samples of the Kubiki test boring in Niigata Prefecture (Iwamatsu, 1971). It is, therefore, reasonable to take layer thicknesses as 'depth' and that the formation of slaty cleavage or shear folding occurs at depths of about 3000 m or more.

Next, we must consider the problem of the lower depth limit at which slaty cleavage will develop. In places in the Kitakami–Abukuma Mountains where it can be established, this depth is about 5000 m. Deeper than this, i.e. up to 7000 m crenulation cleavage and bedding plane schistosity coexist and folds develop with *crenulation cleavage* as the axial-planar cleavage. There is a close relationship in the origins of this crenulation cleavage and the overlying slaty cleavage and they are thought to have formed during the same period. At depths greater than 7000 m one finds flow folding in which bedding plane schistosity has developed, and there are no signs of cleavage. In other words the horizontal distribution of foliated structures in the orogenic belt as mentioned above is believed to have originated in differences of depth inside the orogenic belt. However, as the figures of 5000 and 7000 m incorporate increases in layer thickness due to deformation and apparent reductions in thickness due to the formation of similar folds as described in Section 4.3, they cannot be regarded as definitive. In view of the above information, one may reasonably say that the depth at which slaty cleavage develops is 3000–5000 m (700–1000 bar as confining pressure).

(c)   Structural level and slaty cleavage

As described earlier, a system with a series of folds comprising, from deeper to shallower horizons, flow folding → folding with crenulation cleavage → shear folding → flexural folding, is found in the Kitakami–Abukuma Mountains, Although the the boundaries between styles are gradational, each style is an independent, separate form of folding, and a specific form of folding corresponds to a specific depth. In certain places the deformation may change qualitatively in response to quantitative changes in depth, i.e. temperature and pressure. This is the same type of relationship as is found in the qualitative response of water-containing systems to quantitative changes in the amount of heat, i.e. ice → water → water vapour. In short, a structural sequence is found in folding controlled by depth. This is known as the *structural level of folds* (*faltungsstockwerke*). Similar structural sequences have been reported in orogenic belts throughout the world. For example, the changes of fold styles described above are also observed in areas such as Dutches County in the Appalachians (Maxwell, 1962*), the Thüringen slate mountains in the Hercynians (Schroeder, 1966), the central Pyrenees (Odé, 1966) and the eastern Pyrenees (de Sitter and Zwart, 1960). In each of these cases slaty cleavage developed at middle-range depths in this structural layer.

However, it is far from being the case that such structural sequences will develop completely in all orogenic belts. Even in Japan it is only in the Kitakami–Abukuma mountains that there has been typical development of the above layering. In southwest Japan a sequence composed of flow folding → lens folding → flexural folding is found with very little slaty cleavage (Kimura, 1968*).

Even given deformation at the same middle-range depth, what is different between the places where slaty cleavage develops and where it does not? Tokuyama (1971*) used the results of creep tests for metals and deduced that shear folding was produced at low strain rates, and lens folding at high rates. Ono (1973*) undertook X-ray crystallography studies on slaty cleavage which had formed in the northern area of the Setogawa zone in southwest Japan and discovered that even in the same stratigraphic horizon there was better development of slaty cleavage the higher the temperature of the location. Combining the two together implies that, although there was slow deformation at relatively high temperatures in northeast Japan, deformation in southwest Japan perhaps proceeded relatively rapidly at lower temperatures. No attempts have been made, in research, so far to estimate the tectonic pressure needed for the development of slaty cleavage. This will have to remain as an outstanding topic for future work.

Akira Iwamatsu

References

The literature concerned with rock cleavage is extensive. Siddans (1972*) is particularly recommended.

Beutner, E. D. (1978): Slaty cleavage and related strain in Martinsburg slate, Delaware Water Gap, New Jersey, *Am. J. Sci.*, **278**, 1–23.

Dieterich, J. H. (1969): Origin of cleavage in folded rocks, *Am. J. Sci.*, **267**, 155–65.

Hobbs, B. E., Means, W. D. and Williams, P. F. (1976): *An Outline of Structural Geology*, 571 pp. John Wiley, New York.

Iwamatsu, A. (1969): Structural analysis of the Tsunskizaka syncline, in southern Kitakami mountainous land, Northeast Japan, *Earth Sci. (Chikyu Kagaku)*, **23**, 227–35.

Iwamatsu, A. (1975): Folding-styles and their tectonic levels in the Kitakami and Abukuma mountainous lands, Northeast Japan, *J. Fac. Sci., Univ. Tokyo*, Sec. II, **19**, 95–131.

Kimura, T. (1968): Some folded structures and their distribution in Japan, Japan. *J. Geol. Geogr.*, **39**, 1–26.

Maxwell, J. C. (1962): Origin of slaty and fracture cleavage in the Delaware Water Gap area, New Jersey and Pennsylvania, *Geol. Soc. Am. Mem., Buddington Vol.*, 281–311.

Ono, S. (1973): Slaty cleavage in the Paleogene Setogawa group in Central Japan, *J. Fac. Sci., Univ. Tokyo, Sec. II*, **18**, 431–54.

Ramsay, J. G. (1962): The geometry and mechanics of formation of 'similar' type folds, *J. Geol.*, **70**, 309–27.

Ramsay, J. G. (1967): *Folding and Fracturing of Rocks*, 568 pp. McGraw-Hill, New York.

Siddans, A. W. B. (1972): Slaty cleavage — A review of research since 1815, *Earth Sci. Rev.*, **8**, 205–32.

Tokuyama, A. (1965): Faltungsstockwerke in der Ojika-Halbinsel Nordostijapans, *Sci. Papers Coll. Gen. Educ. Univ. Tokyo*, **15**, 217–36.

Tokuyama, A. (1965): Faltungsstockwerke in der Ojika-Halbinsel Nordostjapans, **77**, 279–87.

Williams, P. F. (1977): Foliation: A review and discussion, *Tectonophys.*, **39**, 305–28.

Wood, D. S. (1974): Current views of the development of slaty cleavage, *Ann. Rev. Earth Planet, Sci.*, **2**, 369–401.

Further Reading

A general summary of the experimental work on slaty cleavage excluded in this volume is:

Means, W. D. (1977): Experimental contributions to the study of foliations in rocks: a review of research since 1960, *Tectonophys.*, **39**, 329–54.

Besides Ramsay, J. G. (1967*) the following are specialist treatises on the structural analysis of deformed rocks:

Turner, F. J. and Weiss, L. E. (1963): *Structural Analysis of Metamorphic Tectonites*, 545 pp. McGraw-Hill, New York.

Whitten, E. H. T. (1966): *Structural Geology of Folded Rocks*, 678 pp. Rand McNally, Chicago.

A photographic account of all types of deformed structures including rock cleavage is given in:

Weiss, L. E. (1972): *The Minor Structures of Deformed Rocks*, 431 pp. Springer, Berlin.

Geological Structures
Edited by T. Uemura and S. Mizutani
©1984 John Wiley & Sons Ltd.

5

Folds and Folding

The process whereby layered structures in geological bodies become bent as a result of deformation is known as 'folding' and the resulting structures as 'folds'.

The earth has been changing ceaselessly throughout all the geological periods, right up to the present day. Folding is just one form of this change. When one encounters strata, presumably deposited horizontally, which have been bent as shown in Figure 5.1, one should be able to feel that folding itself

Figure 5.1   Folds of strata. From the Mesozoic system of the Oga peninsula. (Photo Y. Masai, courtesy of F. Takizawa)

can be quite a considerable change. It is known that large-scale folding, currently in progress, is associated with a continuous change in topography (Sugimura, 1952; Kaizuka, 1972). Such folds are called 'active folds', and many of them are known in the Japanese archipelago, which is located in an active orogenic zone (Research Group for QTM, 1973). When the discussion turns to folds formed in the past, one would immediately think of the grandiose structures in the Swiss Alps where erosion and denudation have been at work to expose the roots of a towering mountain belt. The tight folds shown in Figure 5.2, together with an enormous mountain chain of connecting peaks over 4000 m high, speaks eloquently of the severity of the orogenic movements to which Alps have been subjected. They also demonstrate that folding is a predominant form of deformation in orogenic belts. This is known to be generally true, not only in the Swiss Alps, but also in other old orogenic belts in Japan and in many other parts of the world where the roots of the mountains are exposed. One should be able to decipher the processes of folding in orogenic belts through observations and analyses of those folds (e.g., Figure 5.3).

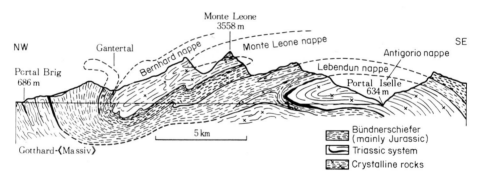

Figure 5.2   Structural profile of Pennine zone, Swiss Alps, drawn along Simplon tunnel (from C. Schmidt, H. Preiswerk, E. Argand and W. Nabbolz in Cadisch, 1953)

It is by no means true to say, however, that folds only develop in orogenic zones. Even under the great plains in the northern parts of Germany, far from the Swiss Alps, the strata are strongly folded. Folding is a common mode of deformation in the earth's crust, and in particular it is one of the principal modes of tectonic movements in orogenic belts. A study of folds and folding is thus one of the most important subjects in structural geology.

Studies of folds, aimed at an understanding of tectonic processes in the earth's crust, have been undertaken along two lines; that is, (1) to examine where in the earth and during what geological periods a certain type of folds formed and to characterize their structural properties, and (2) to classify folds

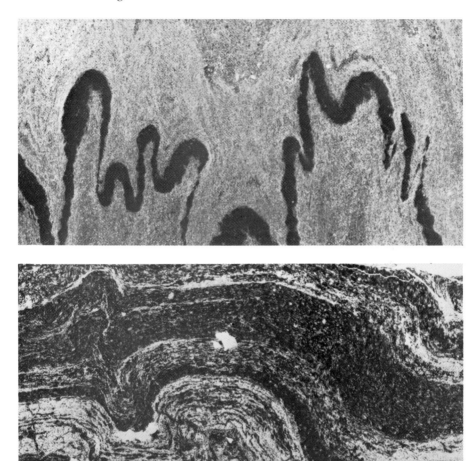

Figure 5.3   Folds of strata. (a) single-layer folds, (b) conjugate folds. From the Sambagawa crystalline schists

and to analyse the mechanisms of folding. Not only descriptions and analyses of natural folds but also folding experiments and theoretical analyses of folding, including model experiments and numerical analyses, have played significant roles. Recent developments in petrofabrics have been providing tools to analyse strain distribution and deformation mechanisms in folded strata. As in many other fields in the earth sciences, studies of folds and folding have been becoming an interdisciplinary science.

This chapter is intended to explain the fundamental concepts which form the basis of modern folding theory, still being developed with contributions from interdisciplinary researches. Basic types of folds are illustrated so as to give some general guidance when observing folds in nature, and the current state of research into their structural features are summarized.

5.1   Fundamental types of folding

If a fold is a deformed structure, then its characteristics should be determined by the mechanical properties of the system and by the external forces applied thereto. A fold manifests itself as a bent, layered structure in a geologic body. This layering is defined by the differences in the types of constituent minerals or in the modal composition, or by textural differences of constituent minerals. In the majority of layered rocks, some layers are more resistant to deformation and are known as '*competent layers*',† whereas others are '*incompetent layers*'† which can be deformed relatively easily. In some cases layering is defined by the parallel arrangement of platy minerals. Platy minerals are usually anisotropic with respect to their mechanical properties, so that the mechanical properties of rocks containing such planar structures also become anisotropic. The interface between two layers is often subjected to inter-layer slip. The mechanical anisotropy of layered rocks may also be produced by such inter-layer slip. All of these properties are important factors in determining the mechanical properties of the system, and significantly control the type of folding.

The external forces which create folds in the crust appear to act in extremely diverse manners. However, research to date indicates that many types of folding can be treated simply (1) as the folding due to compression acting parallel or subparallel to the layering, (2) as the folding due to heterogeneous loading acting at a high angle to the layering, or (3) as the folding caused by the action of gravity. Complex folding can be treated as the superimposition of these basic folding processes. Basic types of folding will be classified in terms of the external forces and their common features will be described briefly in the following.

Consider first the folding of layered rocks subjected to compression parallel or subparallel to the layer. To facilitate an understanding of this type of

---

† Willis (1893) introduced the factor of *competency* as a measure to indicate the tendency of certain layers to press into the surrounding rocks and form folds whilst undergoing bending deformation. He thought that stiff (competent) layers can support external forces while being bent, whereas more deformable (incompetent) layers shorten easily by deformation so that the layers will not exhibit bending deformation to form folds. Even at present, it is not easy to discuss the mechanical meaning of competency in general terms to cover geological bodies with highly diverse mechanical properties. However, many workers since Willis have led to a clear definition of competency in relatively simple fold models. For example, in the case of elastic and viscous layers, the competent layer is the one of two elastic layers with the greater elastic modulus, or the one of two viscous layers with the higher viscosity. The competency of a certain layer is thus determined by the relative differences in mechanical properties between it and the surrounding rocks.

folding, imagine a rod composed of a uniform elastic material and with a regular cross-section. If a force is applied to this rod in its axial direction, it will deform first by uniform contraction. Of course, such homogeneous deformation satisfies the equilibrium equations of forces. However, once the force exceeds a certain limit (critical load), bending deformation of the rod will also occur, to satisfy the equilibrium equations. Then, two or more modes of deformation are possible under this critical load, and no unique solution is established. If one tries to further shorten the rod by uniform deformation, a force proportional to the incremental strain must be applied, resulting in an increase in the potential energy of the system. On the other hand, the potential energy of the system is maintained generally at a low level with bending deformation even if deformation continues. Therefore, once the force has reached the critical load, the uniform deformation can no longer be a stable mode of deformation and a small perturbation readily changes the deformation into bending deformation. Phenomena similar to this are very common in nature and they are treated as problems of instability in mathematical physics. In the mechanics of deformable bodies, the phenomenon in which long thin structures like columns or rods deform by bending when subjected to an axial force is known as '*buckling*'.

Studies of the buckling of building structures were begun over two centuries ago by the mathematicians, L. Euler and J. L. Lagrange. However, it was very long before the correlation between the folding in nature and the buckling was recognized. The work of Willis (1893) was perhaps the most important breakthrough in this respect. He recognized a particular importance of the differences in mechanical properties between individual layers and demonstrated through his model experiments that fold shape is determined primarily by the bending deformation of competent layers, and incompetent layers behave passively adjusting themselves to the deformation of the competent layers. His model was incomplete as a mechanical explanation of folding, but his recognition of the mechanical significance of heterogeneous layering of geologic bodies opened a door for subsequent researches. That the bending deformation of competent layers is nothing but buckling is recognized fully in the theory of Smoluchowski (1909, 1910). It is known at the present time that a certain type of mechanical anisotropy can also cause buckling phenomena (Biot, 1963[*]; he called this '*internal buckling*'). Folds produced by the buckling of competent layers or by internal buckling of anisotropic rocks are called *buckle folds* or *buckling folds*. Buckling as a phenomenon of instablility covers a wide spectrum of deformation behaviours, and the characteristics of buckle folds depend considerably on the internal constitution of the system in regard to the geometry and mechanical properties. The constitution of the system must therefore be specified clearly when discussing buckle folds. Sections 5.2 to 5.6 deal with the characteristics of buckle folds in a few typical systems.

Folds produced by inhomogeneous stresses acting at almost right angles to the layers are known as '*bending folds*'.† In this case one unique mode of deformation corresponds to given external stresses, so that the folding is not an instability phenomenon. Unlike buckle folds, heterogeneous layering or mechanical anisotropy in the system is not a basic prerequisite to the development of bending folds; the layers may simply be any markers to visualize the geometry of folds. The shape of a bending fold is largely controlled by the type of the inhomogeneous external stresses rather than by the internal constitution of the system. Research into this type of folding is relatively recent, dating from the work of Hafner (1951*). Section 5.7 will discuss the characteristics of bending folds.

All materials on and in the earth are subject to the earth's gravitational field and all folding occurs under its influence. Gravity can have completely opposite effects, i.e., a stabilizing or destabilizing effect, in the development of folds. Consider an example in which strata near the surface become folded to produce synclinal and anticlinal undulations. In this case buoyancy is at work in the synclinal section but the anticlinal section is subject to the downward force of its own weight. Growth of the fold is thus hampered by gravity. The stabilizing effect of gravity in this kind of system is generally thought to be more marked as the scale of the fold increases. Thus the gravity is an exceedingly important factor when considering problems of huge geologic structures, over several tens of kilometres across. Stabilization of folding in response to gravity has been investigated mainly at the theoretical level since the work of Smoluchowski (1910).

High-density strata frequently overlie low-density ones in the earth's crust, giving rise to so-called 'inverted density systems'. Examples of geologic bodies containing rock salt are oil or magma. The constitution of such systems is unstable in the gravitational field and the low-density layers tend to rise under their own buoyancy. Since Arrhenius (1912) pointed out that such an inverted density distribution itself is the fundamental cause for the formation of salt domes or diapirs, the upward movements of low-density materials have been examined in detail. More than two rising-modes are possible from the standpoint of continuum mechanics, and in reality the mode with greatest growth-rate is manifested. Therefore, the rise of low-density material under gravity is basically an instability phenomenon.

The effects of gravity on deformation are discussed in general terms by Ramberg (1967*). Sections 5.8 and 5.9 deal with the effects of gravity on the formations of single-layer folds and diapirs, respectively, to illustrate the stabilizing and destabilizing effects of gravity.

---

† Buckling described above is also called '*longitudinal bending*' and the 'bending' used here is also known as '*transverse bending*'.

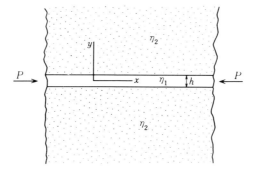

Figure 5.4   Layer of thickness h and viscosity η_1 under compressive stress P in a medium of viscosity η_2

5.2   Single-layer folds

Folds of a single layer enclosed in a medium (Figure 5.4) that form under layer-parallel compression (abbreviated hereafter as single-layer folds) have been studied most extensively either in the analyses of natural folds or in theoretical and experimental analyses of folding. Willis's (1893) concept of competency, as described in the previous section, was the most important breakthrough in grasping the mechanism of folding. Single-layer folds were the first on which detailed theoretical and experimental analyses of folding were performed, based on this idea and including quantitative examination of natural counterparts. Much interesting work has been done since Willis, but Biot's (1957; 1961*) theoretical work marked the real beginning of quantitative studies of single-layer folds. Let us first summarize the findings of his theory of folding whilst recognizing this historical progress.

(a)   Mechanical theory

Consider folding of a system composed of Newtonian fluids. Biot assumed that the folding of a layer could be described by an equation for flexure of a thin beam, that the shape of the fold would be sinusoidal, and that the wavelength of the fold would not change appreciably with deformation. He then theoretically derived the displacement of the layer, w, in the y direction in Figure 5.4 during the progress of deformation as follows (Biot, 1961*):

$$w = w_0 e^{pt} = A_0 e^{pt} \cos lx \qquad (5.1)$$

$$p = P / \left(\frac{4\eta_2}{hl} + \frac{1}{3} \eta_1 h^2 l^2 \right) \tag{5.2}$$

Here, w_0 is the initial value of w, t is the time since deformation began, l is the wave number of the fold ($l = 2\pi/\mathcal{L}$, $\mathcal{L}$ is the fold wavelength), A_0 is the initial value of the fold amplitude, P is the lateral compressive stress, h is the layer thickness, and η_1 and η_2 are the viscosity coefficients for the layer and the medium respectively. The factor, p, indicates the rate of growth of a fold; the larger its value, the faster the fold grows. The first term in the denominator on the right side of equation (5.2) corresponds to the resistance of the medium against the growth of the fold, and the second term to the resistance of the layer itself to bending deformation. The greater the lateral compressive stress, and the lower the viscosity coefficients of the layer and the medium, the faster is the rate of growth of the fold. The above solution also indicates that the growth rate will vary with the fold wavelength for a given lateral compressive stress, P (Figure 5.5). Biot thought that the fold wavelength would be selected progressively during deformation so as to attain the value of the greatest growth rate, i.e., the value that maximizes p in equation (5.2). He called this '*dominant wavelength*' and derived from equation (5.2):

$$\frac{\mathcal{L}_d}{h} = 2\pi \sqrt[3]{\frac{\eta_1}{6\eta_2}} \tag{5.3}$$

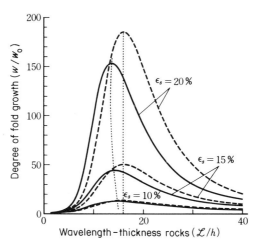

Figure 5.5   Relationship between wavelength–thickness ratio $\mathcal{L}/h$ and degree of fold growth w/w_0 (= amplification of amplitude) for single-layer fold obtained by Biot's theory (1961*, 1965b*). Dashed lines and solid lines show degrees of fold growth obtained by equations (5.1) and (5.4). Dotted lines connect points where w/w_0 is maximized. ϵ_s is the average strain of the system expressed as the logarithmic strain

Biot (1961*) performed a detailed analysis of the wavelength selection on the basis of equations (5.1) and (5.2) and showed that the greater the viscosity ratio, η_1/η_2, the more regular wavelengths would the folds attain.

Biot (1959, 1965a*) and Biot and Ode (1962) solved the problem more precisely and showed that the theory based on the equation for the flexure of a thin beam holds reasonably well when $\mathcal{L}_d/h > 10$ and that the adhesion between the layer and the medium does not alter the dominant wavelength significantly. Biot's theory is formulated on the basis of the viscoelastic correspondence principle† and the folding of systems with various mechanical properties can be treated in a unified manner.

The thickness of the layer increases due to lateral compressive stress and the fold wavelength decreases during the folding, but these points were ignored in Biot's theory above. Biot (1964, 1965b*) then developed a theory which took into account the uniform shortening of the layer, and derived:‡

$$\frac{W}{W_0} = A(\epsilon_s)e^{\epsilon_s} \tag{5.4}$$

$$\text{In } A(\epsilon_s) \simeq \frac{4(1-c)\gamma_{av}\epsilon_s}{(1 + c^2)\sinh 2\gamma_{av} + 2c \cosh 2\gamma_{av} + 2(c^2 - 1)\gamma_{av}} \tag{5.5}$$

$$c = \eta_2/\eta_1, \qquad \gamma_{av} = (\pi h_0/\mathcal{L}_0)e^{\epsilon_s} \tag{5.6}$$

where ϵ_s is the average strain (average shortening) of the system expressed as the logarithmic or natural strain, and $\mathcal{L}_0$ and h_0 are the initial values for the wavelength and the layer thickness, respectively.

Equation (5.4) is shown as the solid lines in Figures 5.5. Note that the dominant wavelength changes with the progress of deformation. The dominant wavelength–thickness ratio, L_d/h, maximizes $A(\epsilon_s)$ and is expressed as:

$$\frac{2\eta_2/\eta_1}{1 + (\eta_2/\eta_1)^2} = \frac{2\gamma_d - \tanh 2\gamma_d}{1 - 2\gamma_d \tanh 2\gamma_d} \tag{5.7}$$

† Based on this principle, an elastic solution can be extended to a visco elastic solution using the operational or Laplace transform method (Biot, 1954, 1965*; Pipkin, 1972; Schapery, 1964*, 1972). Schapery's (1964*) correspondence principle can also be applied to non-linear constitutive equations describing the flow of rocks under high temperatures and pressures (e.g. Weertman, 1970; Stocker and Ashby, 1973), so that the theories of folding could be extended relatively easily to systems with such mechanical properties.

‡ These equations are obtained from Biot's results when transformation of $\epsilon_s = (P/4\eta)t$ is undertaken. The average logarithmic strain of the system ϵ_s is used instead of time t in equations (5.4) to (5.9), because ϵ_s can be estimated more easily for natural folds than t. See Ramsay (1967*, p. 53) for the definition of logarithmic strain.

$$\gamma_d = \pi h_0/\mathcal{L}_d = (\pi h/\mathcal{L}_d)e^{-\epsilon_s} \tag{5.8}$$

For large values of η_1/η_2 and $\mathcal{L}/h$, $c \ll 1$ and $\gamma_{av} \ll 1$, and $A(\epsilon_s)$ in (5.5) is at maximum when $\mathcal{L}_d/h$ is given by:

$$\frac{\mathcal{L}_d}{h_0} = \frac{\mathcal{L}_d}{h}e^{\epsilon_s} = 2\pi \sqrt[3]{\frac{\eta_1}{6\eta_2}} \tag{5.9}$$

Since ϵ_s is contained in these equations, ϵ_s must be evaluated in order to estimate η_1/η_2 from the measurements of $\mathcal{L}_d/h$ for natural folds. Sherwin and Chapple (1968) also modified Biot's theory, taking into account the uniform shortening of the layer and derived an equation for the dominant wavelength. Their equation gives almost the same result as does equation (5.9).

The degree of fold growth is indicated by w/w_0 in equation (5.4), which is the product of $A(\epsilon_s)$ due to folding and (ϵ_s) due to the uniform shortening of the system. Examination of equations (5.4) and (5.5) reveals that the growth of fold is initially slow, but the growth rate increases exponentially with the progress of deformation. This rapid growth of fold corresponds to the rapid increase in the term, $A(\epsilon_s)$. Although uniform shortening of the layer is the principal form of deformation in the early stages of deformation, heterogeneous deformation associated with folding becomes progressively predominant. Ramberg (1964)[*] called the former *'layer-parallel shortening'* and the latter *'buckle shortening'*.

(b)   The existence of the dominant wavelength

Tens of years before Biot's theory was published, Willis and Willis (1934) had recognized an empirical relationship: 'the thicker and more competent the layer, the longer the wavelength of the fold'. This agrees qualitatively with equation (5.3) which gives the dominant wavelength. Unfortunately, however, they conducted no quantitative analysis of the fold wavelengths. The relationship between the fold wavelength and the layer thickness for natural folds was first examined quantitatively by Sherwin and Chapple (1968), Ramberg and Ghosh (1968) and Hara *et al.* (1968[*]). They showed that the fold arc–length–thickness[†] ratio, $\mathcal{L}_d/h$, varied with the combination of rocks comprising the layer and the medium, and $\mathcal{L}_d/h$ possessed relatively low values, mostly less than 10.

[†] The positions of the fold axes are fixed at a relatively early stage of fold growth, and wavelength selection effectively ceases thereafter. Subsequently the layer-parallel shortening is markedly less than the buckle shortening, and so the arc-length of the fold remains fairly constant. On the other hand, the fold wavelength continues to decrease with the growth of the fold. Therefore, in order to determine the dominant wavelength for real folds at various stages of their growth, it is more convenient to use the fold arc-length rather than the wavelength. Assuming a sinusoidal fold shape, the dominant wavelength of a fold at a certain stage of its growth can be estimated easily from this dominant arc-length.

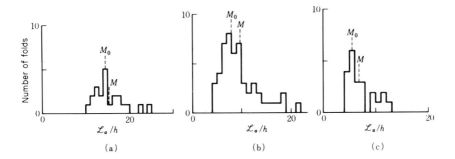

Figure 5.6 Frequency histogram of arc length–thickness ratio $\mathscr{L}_a/h$ for single-layer folds in the Sambagawa crystalline schists. (a) folds of quartz veins enclosed in pelitic schist (Hara *et al.*, 1968*), (b) folds of quartz veins enclosed in psammitic schist (Hara, 1974), (c) folds of quartz veins enclosed in basic schist (Hara *et al.*, 1973*). M_0: mode value, M: mean value

Figure 5.6 shows frequency distribution diagrams of the measured $\mathscr{L}_a/h$ for single-layer folds of the Sambagawa crystalline schists. The frequency distribution of $\mathscr{L}_a/h$ is skewed towards its smaller value. The shape of this distribution curve resembles that of a curve showing the theoretical relationship between w/w_0 and $\mathscr{L}/h$ (cf. Figures 5.5 and 5.6). Therefore, the dominant wavelength in the theory should be identified as the mode of measured $\mathscr{L}_a/h$. Measurements of the dominant wavelength are extremely important in that they enable one to deduce the viscosity ratio (or more generally, differences in competency between layers or their relative mechanical properties) between the layer and its medium at the time of folding. The dominant wavelength varies with the progress of deformation, so that one must estimate the average strain of the system and the wavelength–thickness ratio at a stage when the position of the fold axis is fixed and the process of wavelength selection effectively ceases, in order to calculate the viscosity ratio from the wavelength–thickness ratio. Hudleston (1973*) established in his model experiments that wavelength selection ceases when the angle between the two limbs of a fold (i.e. the interlimb angle, θ) reaches 130 – 150°. If this result is general, one must estimate $\mathscr{L}_a/h$ and ϵ_s corresponding to $\theta = 130 – 150°$ from their measured values. Shimamoto and Hara (1976*) estimated $\mathscr{L}_a/h$ ϵ_s at $\theta = 150°$ for single-layer folds of quartz veins in the Sambagawa crystalline schists, and calculated the viscosity ratio between them using equations (5.7) and (5.9) as shown in Table 5.1.

Equation (5.7) gives slightly lower viscosity ratio than does equation (5.9). Considering the approximation involved in the theory, equation (5.7) should yield more accurate results. It can be seen from Table 5.1 that the viscosity decreases in the order of quartz veins, basic schists, psammitic schists and

Table 5.1   Estimation of viscosity ratio η_1/η_2 between quartz veins and surrounding Sambagawa schists based on arc length–thickness ratio $\mathscr{L}_a/h$ for single-layer folds of quartz veins (Shimamoto and Hara, 1976*)

Composition of system (corresponding to a, b and c of Figure 5.6)	$\mathscr{L}_a/h$ mode value	η_1/η_2 based on equation (5.7)	η_1/η_2 based on equation (5.9)
(a) Quartz vein/pelitic schist	14.5	94–136	103–147
(b) Quartz vein/psammitic schist	8.0	23–33	29–40
(c) Quartz vein/basic schist	5.5	14–26	19–32

Note: As the average strain of the system is not precisely determined at the stage when wavelength selection ceases, η_1/η_2 is not directly determined. Here η_1/η_2 is estimated allowing for the anticipated range of error.

pelitic schists at the time of Hijikawa-phase† deformation (Hara *et al.*, 1977). The viscosity ratio between the most competent quartz veins and the most incompetent pelitic schists is about 115.

Calcite veins in pelitic schists in the Sambagawa belt are folded with arc–length–thickness ratios of about 5. Judging from the results in Table 5.1, the viscosity ratio between quartz and calcite veins is perhaps on the order of 5–10. Parrish *et al.* (1976*) recently show that the effective viscosity ratio between quartzite and marble varies considerably with temperature under wet conditions. Assuming a typical geological strain rate of 10^{-14}/s for the deformation associated with the formation of these folds, temperature at the time of folding is estimated to have been slightly less than 400°C from the inferred viscosity ratio and the experimental results of Parrish *et al.* (1976*). If in the future constitutive equations are to be established for a variety of rocks, one should be able to estimate more precisely the physical conditions at the time of folding.

(c)   Strain distributions and internal structures

Various deformed structures develop in and around folded layers, reflecting the stress and strain distributions developed in the system during the folding. The features of the internal structures vary with the type of folding, and hence their analyses would lead to the understanding of how folds formed. The

† Deformation phase following the major phase of the recrystallization of metamorphic minerals; its geological period is estimated as the late Jurassic to early Cretaceous. Number of large- and small-scale folds formed throughout the Sambagawa belt during this phase, including anticlines and synclines with their axes aligned in an *en échelon* arrangement, slightly oblique to the trend of the belt, and minor folds whose axial surfaces are at high angles to the schistosity.

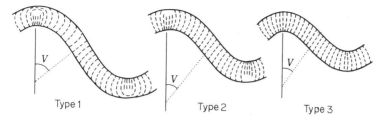

Figure 5.7   Orientation patterns of schistosity or cleavage in competent layers forming single-layer folds (Hara *et al.*, 1968*). Dashed lines show orientation of schistosity or cleavage. *V* is the angle between the axial surface of fold and the orientation of schistosity or cleavage at the inflection point

characteristics of the internal structures should also change with the amount of the average strain of the system, the mechanical properties of the constituent rocks, and the temperature and lithostatic and pore water pressures at the time of folding. This section only deals with the strain distribution and related structures in and around single-layer folds. For a layer with no layering inside, the internal structures that develop during the folding are commonly schistosity or slaty cleavage. The significance of the schistosity or cleavage pattern in the understanding of the mechanism of folding was recognized only recently, and the pattern has been studied with the view toward a detailed analysis of the strain distribution (see Chapter 4.) Cloos's (1947) work using ooids on the strain analysis of the South Mountain folds in Maryland is perhaps the first extensive study along this line. Strain distributions and the internal structures of single-layer folds were examined comprehensively by Hara *et al.* (1968*), who analysed single-layer folds of quartz veins in a medium of Sambagawa crystalline schists. They recognized three basic types of schistosity or cleavage patterns within folded single-layers as shown in Figure 5.7. In all of these types, the schistosity or cleavage develops in a fan-like arrangement converging towards the fold core. The major difference between the types is seen in the outer edge of the axial zone. In type I, the schistosity or cleavage is parallel to the layer in the outermost edge, and there is a zone called neutral surface where schistosity or cleavage is not developed. In type II the neutral surface is located on the outermost edge, and in type III no neutral surface is developed within the layer. Another difference between the three types can be seen in the angle, *V*, between the axial surface of the fold and the schistosity or cleavage at the inflection point. For the same interlimb angle, *V* decreases from type I to type III (Figure 5.7). The distinction between the three types is of course arbitrary, and there is a continuous change from one type to another. For instance, the neutral surface in type I may be located anywhere between the centre of the layer and its outermost edge, and the angle, *V*, in type III varies continuously from that shown for type II in Figure 5.7 down to virtually zero in some cases.

Hara *et al.* (1968*, 1973*) correlated these arrangements of schistosity or cleavage with the data on arc–length–thickness ratio, and found that the type of schistosity or cleavage pattern in a folded layer depends mainly on the competency difference between the layer and its medium (i.e. viscosity ratio in Table 5.1). The recognized types I to III patterns for folded quartz veins in pelitic schists, types II and III patterns for those in psammitic schists, and type III pattern for those in mafic schists. This combined with the results in Table 5.1 clearly shows that the viscosity ratio must be large if a fold is to possess the type I schistosity or cleavage pattern, and that if it is small type III tends to show up. As discussed in the above, the smaller the viscosity ratio, the greater the effect of uniform layer shortening. Thus, Hara *et al.* (1968*) argued that the layer-parallel extension at the outer edge due to the bending deformation becomes predominated by the uniform layer-parallel shortening when the viscosity ratio is small, resulting in the development of type III pattern. These observations and inference were confirmed by subsequent experiments as shown below. Schistosity and cleavage are deformed structures that develop perpendicular to the direction of the maximum compressive strain (see Chapter 4). Thus, the types of schistosity or cleavage patterns can be identified as those of strain distribution.

Dieterich (1969*) analysed strain patterns in single-layer folds using a finite element method and reported types II and III patterns. Shimamoto and Hara (1976*) later used the same method to examine strain patterns in still more detail and recognized all three types. They found that the patterns varied not just with the viscosity ratio (or competency difference) but also depended on the initial deflection of the layer and on the average strain of the system (Figure 5.8). For a given average strain of the system, the larger the viscosity ratio and the initial displacement, the deeper will be the position of the neutral surface if the strain pattern is of type I, and the greater will be the angle, V, shown in Figure 5.7. Therefore, the strain pattern in a group of folds formed in a system with a given viscosity ratio will vary reflecting the diversity of the initial deflection of layers. However, it is also clear that types I and III strain patterns tend to appear in systems with large and small viscosity ratios, respectively. The pattern of schistosity or cleavage in the incompetent medium has been classified into three types as shown in Figure 5.9 by Hara (1966*) and Roberts (1971*, 1972). In all types, schistosity or cleavage is arranged in a slightly opened fan. The major difference between them is seen as follows in the region adjacent to the outer edge of the axial zone of the competent layer. In type A, schistosity or cleavage is parallel to the layer immediately adjacent to the layer, next to this there is a spot called neutral axis where no schistosity or cleavage is formed, and still further away it is parallel to the axial surface of the fold. In type B the neutral axis is located immediately adjacent to the outer edge of the folded layer, and in type C no neutral axis is developed in the medium. The position of the neutral axis

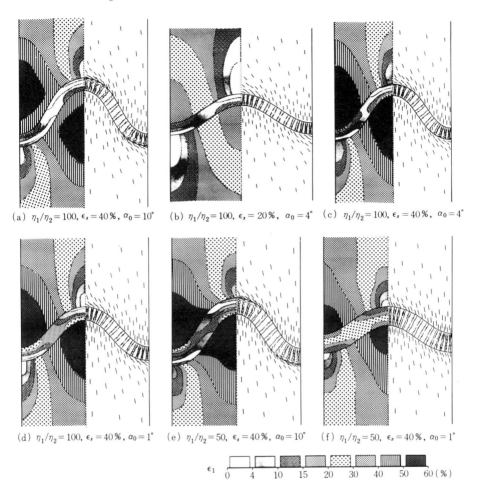

(a) $\eta_1/\eta_2 = 100$, $\epsilon_s = 40\%$, $\alpha_0 = 10°$ (b) $\eta_1/\eta_2 = 100$, $\epsilon_s = 20\%$, $\alpha_0 = 4°$ (c) $\eta_1/\eta_2 = 100$, $\epsilon_s = 40\%$, $\alpha_0 = 4°$

(d) $\eta_1/\eta_2 = 100$, $\epsilon_s = 40\%$, $\alpha_0 = 1°$ (e) $\eta_1/\eta_2 = 50$, $\epsilon_s = 40\%$, $\alpha_0 = 10°$ (f) $\eta_1/\eta_2 = 50$, $\epsilon_s = 40\%$, $\alpha_0 = 1°$

ϵ_1 0 4 10 15 20 30 40 50 60 (%)

Figure 5.8   Strain patterns in single-layer folds obtained by finite-element method (Shimamoto and Hara, 1976*). Left half and right half of each diagram show magnitudes and orientations of the maximum principal elongation ϵ_1 (elongation counted as positive). η_1 +/η_2: viscosity ratio, ϵ_s: average strain of the system. α_0: initial limb dip

which is a factor governing the pattern of the schistosity or cleavage may be found in various positions on the axial surface of the fold in type A. Patterns of schistosity or cleavage in incompetent medium adjacent to the inner side of the axial zone are almost the same in the three types, and it is arranged in a slightly open fan converging towards the inner side.

Strain pattern in incompetent medium adjacent to the outer edge of the

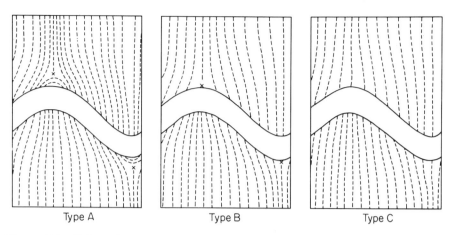

Type A   Type B   Type C

Figure 5.9   Orientation patterns of schistosity or cleavage in incompetent layers surrounding single-layer fold. Dashed lines show orientation of schistosity or cleavage. x is neutral axis of strain

axial zone is determined by a superposition of uniform shortening parallel to the layer and the heterogeneous deformation produced by the growth of the fold pushing into the incompetent medium (the latter being equivalent to the deformation associated with a transverse bending fold; see Section 5.7 below). The growth rate of a fold, as described earlier, increases with increasing viscosity ratio and initial deflection. Therefore, the type A pattern shows up in folds with large initial deflections and with high viscosity ratios, and the type C in folds with small initial deflections and with low viscosity ratios. This is confirmed by the finite element analyses of Dieterich (1969*), Parrish (1973) and Shimamoto and Hara (1976) and by the photoelastic experiments of Roberts and Strömgård (1972*). The results of Shimamoto and Hara are shown in Figure 5.8. It should be remarked that strain patterns in incompetent medium around the outer edge of the axial zone possess the same features as those of transverse bending folds produced under the simultaneous action of lateral compressive stress.

(d)   Variation in layer thickness

The thickness of folded layers is measured on a plane perpendicular to the fold axis. In practice, however, one is often at a loss as to how to measure the thickness of folded layers. One must therefore state clearly which method has been used when measurements of the thickness of a layer have been made. The method of measurement should preferably be applicable to any form of fold and should be clearly able to demonstrate variation in layer thickness throughout it. In view of these points, Ramsay's methods (1967*), orthogonal thickness and thickness parallel to axial surface (axial plane thickness) (Figure A.8 in the Appendix), appear to be useful. Ramsay classified types of folds on the basis of variations in layer thickness. His methods will be summarized below.

Orthogonal thickness is described as the thickness t_β measured from tangents drawn parallel to the upper and lower edge surfaces of a folded layer. The measuring position is shown by the angle β which the tangent at the position forms with the normal to the axial surface. The fold hinge is the position where $\beta = 0$. The line joining points on the folded surfaces where the tangents to the surfaces make an angle of β is called dip isogon. The features of the variation in layer thickness throughout the fold are demonstrated by recording the proportional thickness t_β with the thickness t_0 measured at the fold hinge, $t_\beta' = t_\beta/t_0$, and constructing a diagram showing relationship between t_β' and β such as shown in Figure 5.10.

Axial plane thickness is described as the thickness T_β for a folded layer measured in a direction parallel to the axial surface. The measuring position is shown by measuring β at that position. At that time one must record on which surface of the layer β was measured. The thickness T_β is recorded as a

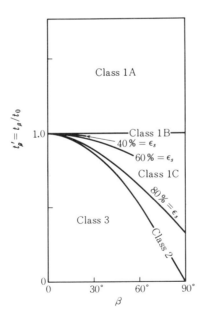

Figure 5.10 Experimental results showing variations in orthogonal thickness t_β' ($= t_\beta/t_0$) during folding processes (Hudleston and Stephansson, 1973*). ϵ_s: average strain of the system, viscosity ratio for layer-medium $= 10$, initial value of wavelength–thickness ratio $= 7.4$

proportion of the thickness T_0 (= t_0) measured at the fold hinge, i.e. $T'_\beta = T_\beta/T_0$, and the variation in layer thickness, throughout the fold can also be illustrated by constructing a diagram showing relationship between $T'\beta$ and β.

In folds in which there is no variation in the layer thickness T_β, i.e. $T'_\beta = 1$, the upper and lower surfaces of the layer show the same geometrical form and the dip isogons constantly run parallel to the axial surface. Folds showing layer thickness variation of this nature are called Class 2 folds, but this is the type of folds which have been so far known as *similar folds*. Layer thickness variation such that $T'_\beta > 1$ is observed in folds with the dip isogons tending to converge towards the fold core. This type of fold is described as Class 1. T'_β is less than 1 ($T'_\beta < 1$) in folds in which the dip isogons converge towards the side on the outer edge of the axial zone. These are called Class 3 folds. Class 1 folds can further be classified by the manner of variation in t'_β (Figure 5.10). Folds in which $t'_\beta = 1$ are called Class 1B but correspond to those which have been so far known as *parallel folds*. Folds in which the layer thickness t_β increases towards the inflection point are called Class 1A, whereas folds in which the layer thickness t_β decreases towards the inflection point Class 1C.

Van Hise (1896) who studied the geology of the Appalachian mountains claimed that Class 1B (parallel) folds and Class 2 (similar) folds predominantly occur. This emphasis has been supported by many workers who studied the geology of many other districts. Parallel folds have been considered to be characteristic of competent layers and similar folds of incompetent layers. Theories about the formation mechanisms of folds up to about the end of the 1950s aimed at explaining these two types of layer thickness variation. A mechanism proposed to explain *parallel folds* was flexural-slip, parallel to layer surfaces, whereas shear folding, based on non-uniform slippage parallel to the axial surface, was suggested to explain *similar folds*. However, the strain and schistosity or cleavage patterns in folds explained in Section 5.2(c) clearly show that these two types of folding mechanisms are not available.

By the way, do the layer thickness variations of single-layer folds generally belong to Class 1B? The variation in layer thickness throughout the fold is the combined total of initial variation and that induced by the folding. Thickness variations induced by the folding are usually to be estimated indirectly from strain analysis of the fold. Assuming that the shapes of quartz grains in single-layer folds (interlimb angle $\simeq 65°$) of quartz veins in the Sambagawa pelitic and psammitic schists approximate to strain ellipsoid, Hara *et al.* (1968*) estimated the layer-parallel compressive strain in each position of the folded layer which was produced by the folding (Figure 5.11). This compressive strain corresponds with the increase in layer thickness. Figure 5.11 therefore demonstrates a relationship between the strain patterns (types I, II and III) and the variation in layer thickness. The phenomena common to the three types of strain pattern are that there is an increase in layer thickness at

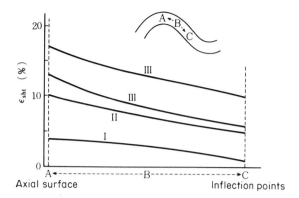

Figure 5.11   Distribution of the amount of layer-parallel shortening (ϵ_{sht}) in the folds (Hara *et al.*, 1968*). I, II and III correspond to the orientation patterns of schistosity or cleavage shown in Figure 5.7

all positions and that the magnitude of increase in layer thickness increases from the inflection point towards the axial zone and the layer thickness variations throughout the folds are of Class 1C. The magnitude of thickness variation from the axial zone towards the inflection point is small in type I but increases towards types II and III. In other words, type I shows features of thickness variation most reminiscent of Class 1B, and type III the most unlike Class 1B. Strain patterns of type III are the most common in natural folds. One may therefore say that the Class 1B folds expected by most workers since van Hise hardly develop at all in single-layer buckle folds. The characteristics of thickness variations may also vary with the process of folding. Figure 5.10 illustrates the experimental results of Hudleston and Stephansson (1973*) for viscous buckle folding using the finite element method. As folding proceeds the characteristics of layer thickness variations may be seen to shift gradually from being close to Class 1B to being close to Class 2.

5.3   Intrafolial folds

Intrafolial folds such as those shown in Figure 5.12 are frequently observed within crystalline schists and gneisses which exhibit well-developed schistosity and/or banded structures, in alternated sandstone and shale, and in banded cherts. Intrafolial folds are characterized by the fold amplitudes which diminish upwards and downwards as shown in the figure. In other words, folded layers gradually change into non-folded ones, and the folded portion is

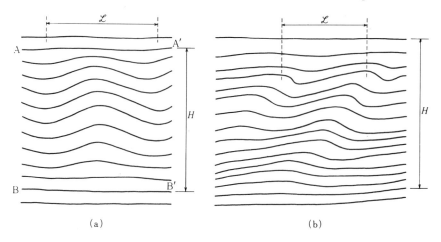

(a)   (b)

Figure 5.12 Schematic forms of intrafolial folds. (a) Folds with orthorhombic symmetry, (b) folds with monoclinic symmetry, $\mathscr{L}$: wavelength of fold, H: distance between non-folded layers (AA', BB') by which the folded portion is confined

confined by flat layers (AA' and BB' in Figure 5.12). One of the most significant features of intrafolial folds is that there is commonly no marked difference in the constitution (i.e. mineral composition and texture) of layers between folded and non-folded portions; that is, the fold forms in a system with uniform constitution. Therefore, one cannot seek the causative factor for this type of folding in the competency difference between the layer and its medium, as in the case of single-layer folds.

The origin of intrafolial folds has been left untouched until fairly recently. Biot (1963*) showed that a certain type of mechanical anisotropy could be a causative factor of folding (theory of internal buckling), and his theory was an important breakthrough in understanding the mechanism of intrafolial folding. Let us begin with briefly summarizing his theoretical results, examine the geometry and strain distribution of natural intrafolial folds, and finally discuss the origin of intrafolial folds.

(a)   Internal buckling of anisotropic elastic bodies

Biot (1963*) considered a system with internal layering (e.g. schistosity, alternating layers with different mechanical properties) under the initial compressive stress, P, parallel to the layering. Taking x and y co-ordinates as parallel to P and as perpendicular to the layering, respectively, he analysed the folding as a plane-strain problem in the x–y plane. Biot assumed homogeneous deformation of the system under the initial stress, P, and dealt with the infinitesimal heterogeneous deformation superposed on this uniform initial strain. That is, his basic premises are that folding occurs when this infinitesimal deformation grows with only small perturbations and that the nature of this heterogeneous deformation determines the shape of the fold.

Biot assumed incompressibility of the material for the sake of simplicity, and analysed the deformation of an anisotropic elastic body whose incremental stress–strain relation is given by:

$$s_{xx} - s_{yy} = 4Ne_{xx} = -4Ne_{yy}, \qquad s_{xy} = 2Qe_{xy} \qquad (5.10)$$

Here, s_{xx}, s_{yy} and s_{xy} are incremental stresses; e_{xx}, e_{yy} and e_{xy} are incremental strains; and N and Q are the elastic moduli corresponding to pure-shear and simple-shear incremental deformations, respectively. The mechanical properties of the rock are assumed to be invariant in the planes parallel to the layering so the equation (5.10) does not change with the choice of the x axis in this plane.

For a system composed of two, regularly alternated, types of layers with different mechanical properties, equation (5.10) gives the average mechanical property of the system. If these layers are mechanically isotropic and if no slippage occurs between them, N and Q are given by (Biot, 1963*, 1965a*):

$$N = \mu_1\alpha_1 + \mu_2\alpha_2$$

$$1/Q = \alpha_1/\mu_1 + \alpha_2/\mu_2 \qquad (5.11)$$

where μ_1 and μ_2 are the elastic moduli for the two types of layers ($\mu_1 \geqq \mu_2$), and α_1 and α_2 are the fractions of the thicknesses of the layers ($\alpha_1 + \alpha_2 = 1$). From equation (5.11) one obtains:

$$N-Q = \frac{\alpha_1\alpha_2(\mu_1 - \mu_2)^2}{\alpha_1\mu_2 + \alpha_2\mu_1} \qquad (5.12)$$

Note that unless μ_1 and μ_2 are equal, N is greater than Q ($N > Q$), and the system as a whole behaves like a mechanically anisotropic body. Let us consider the case in which incremental displacement, u and v, in the x and y directions are expressed only in terms of a scalar function ϕ as:

$$u = -\frac{\partial \phi}{\partial y}, \qquad v = \frac{\partial \phi}{\partial x} \qquad (5.13)$$

This displacement field automatically satisfies the condition of incompressibility. It is known from the theory of incremental deformation that ϕ satisfies a fourth-order linear partial differential equation (Biot, 1965a*). The phenomenon of internal buckling is closely related to the fact that this differential equation possesses a hyperbolic solution. In this case ϕ may be written as follows using a certain function, φ, and a real parameter, ξ:

$$\phi = \varphi(x - \xi y) \qquad (5.14)$$

An important property of equation (5.14) is that ϕ does not change as long as $x - \xi y$ maintains a certain constant value, k. Because incremental displacements are determined solely by ϕ, the displacement field is the same along a line given by $x - \xi y = k$. The direction along which this deformation pattern remains unchanged is called '*characteristic direction*', with its slope given by ξ. Biot did not analyse dynamic aspects of deformation, but physically, a solution in the form of equation (5.14) should correspond to a phenomenon in which local deformation is propagated along the characteristic direction. Biot called the phenomenon in which uniform deformation changes into heterogeneous deformation in the form of equation (5.14) '*internal buckling*'. Let new elastic moduli, M and L, including the initial stress, P, be defined by:

$$M = N + P/4 \qquad L = Q + P/2 \qquad (5.15)$$

Then the internal buckling will occur when P reaches the critical value given by:

$$P = L\xi^4 + 2(2M-L)\xi^2 + L \qquad (5.16)$$

Biot distinguished two types of internal buckling, and called the internal buckling satisfying:

$$2M > L, \qquad P > L \qquad (5.17)$$

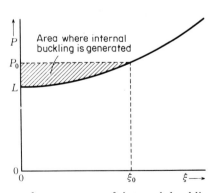

Figure 5.13 Conditions of occurrence of internal buckling of the first kind in anisotropic elastic bodies (Biot, 1963*). The curve schematically represents equation (5.16)

'*internal buckling of the first kind*' (see Section 5.4 for internal buckling of the second kind). When equation (5.17) holds, P has the minimum value of L at $\xi = 0$ (Figure 5.13), and therefore $\xi = 0$ is the most unstable characteristic direction. Equation (5.16) has two real roots $\pm\xi_0$ if $P = P_0 > L$.

The displacement field given by equation (5.14) depends totally on the choice of the function, φ. The selection of φ must therefore be examined for each problem at hand, and the physical meaning of ξ changes accordingly. If one takes the cosine function as a particular solution of equation (5.14), then:

$$\phi = -\frac{1}{2}C\{\cos l(x-\xi y) + \cos l(x + \xi y)\} = -C \cos lx \cos l\xi y \quad (5.18)$$

which on substitution into equation (5.13) yields:

$$u = -\partial\phi/\partial y = -Cl\xi \cos lx \sin \xi ly$$

$$v = \partial\phi/\partial x = Cl \sin lx \cos \xi ly \quad (5.19)$$

Here, C is an arbitrary constant indicating the degree of fold growth, and l is the wavenumber in the x direction ($= 2\pi/\mathscr{L}$, $\mathscr{L}$ is the fold wavelength). The displacement field given by equation (5.19) specifies a geometry of deformed structure similar to that of intrafolial folds that are symmetrical with respect to their axial surface. Thus, Biot's theory accounts for the geometry of intrafolial folds reasonably well. Comparing the displacement field, equation (5.19), with the intrafolial folds in Figure 5.12(a), one would obtain ξ in terms of the fold wavelength, $\mathscr{L}$, and the thickness of the folded section, H, as:

$$\xi = \mathscr{L}/2H \quad (5.20)$$

It can be seen from Figure 5.13 that the lower the value of ξ, i.e. the greater the thickness, H, relative to the wavelength, $\mathscr{L}$, the smaller will be the initial compressive stress, P, required for the onset of internal buckling.

(b) Internal buckling of anisotropic viscous fluids

Geological deformation during folding is mostly ductile. Therefore viscous models rather than elastic ones should be more realistic in the simulation of natural folding. Biot (1965a*) applied fluid mechanics to extend the above theory to include internal buckling of anisotropic viscous fluids. Major results from his analysis will now be summarized below.

Assuming incompressibility of materials as in the elastic case, the constitutive equation for anisotropic viscous fluids would be written in the same form as equation (5.10) with N and Q being replaced by the viscosity coefficients,

η_n and η_t, and with the strain components by corresponding strain rate components, respectively. For a system composed of two types of regularly alternated layers with viscosity coefficients, η_1 and η_2, equations for η_n and η_t are identical in form to equation (5.11) (replace μ_1 and μ_2 with η_1 and η_2, respectively).

There are two principal differences between the elastic and viscous solutions. First, there is no critical load for the initiation of viscous folding. In other words, the fold will grow no matter how small the compressive stress, P, is. Secondly, although the quasistatic elastic solution can only predict the general form of a fold, the process of folding is analysed in the viscous solution.

Viscous layers subjected to the compressive stress, P, undergo not only heterogeneous deformation associated with the folding but also uniform layer-parallel shortening. Assuming the AA' and BB' in Figure 5.12(a) remain flat during the folding, changes in $\mathscr{L}$, H, l and ξ in equation (5.19) due to this uniform shortening are express as:

$$\mathscr{L} = \mathscr{L}_0 e^{-\epsilon_s}$$

$$H = H_0 e^{\epsilon_s}$$

$$l = l_0 e^{\epsilon_s}$$

$$\xi = \mathscr{L}/2H = \xi_0 e^{-2\epsilon_s} \tag{5.21}$$

where ϵ_s is the average shortening of the system expressed as the logarithmic strain, and the quantities on the right-hand sides represent their initial values. Displacement field in the viscous case can also be written in the same form as equation (5.19), with Cl representing the maximum fold-amplitude in the system. Let Cl at the average strain of ϵ_s be denoted by w and its initial value by w_0. Then the amplificiation factor, w/w_0 (a parameter indicating the degree of fold growth), is given approximately by (Biot, 1965c*; t in his equations is replaced here by ϵ_s):

$$w/w_0 = A(\epsilon_s)e^{\epsilon_s} \tag{5.22}$$

$$\ln A(\epsilon_s) \simeq 4\left(\frac{\eta_n}{\eta_t} - 1\right) \frac{(1 - \xi_{av}^2)\epsilon_s}{1 + 2\left(2\frac{\eta_n}{\eta_t} - 1\right)\xi_{av}^2 + \xi_{av}^4} \tag{5.23}$$

$$\xi_{av} = \xi_0 e^{-\epsilon_s} \tag{5.24}$$

This result demonstrates that the larger the viscosity ratio between the two types of alternating layers (or more generally, the larger the degree of the

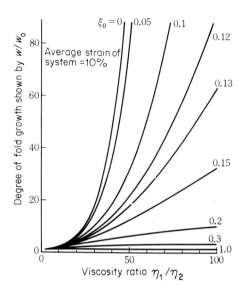

Figure 5.14 Growth process of intrafolial folds in layers of equal thickness and alternate viscosity coefficients η_1 and η_2 (Biot, 1965*). This represents relationship between degree of fold growth w/w_0 and viscosity ratio η_1/η_2 for various values of ξ_0 which is obtained by equation (5.22)

mechanical anisotropy of the system), or the smaller the value of ξ_0 (the greater H_0 relative to to $\mathcal{L}_0$), the more rapid will be the fold growth (Figure 5.14). This conclusion does not differ fundamentally from the characteristic of the elastic solution.

(c) Finite element analysis of intrafolial folds

Biot's theory can be supplemented by a finite element analysis of folding in that finite deformation and stress and strain distributions in the system, can be analysed with reasonable precision. Figure 5.15 is a representative example of intrafolial folds of multilayered viscous fluids analysed by Shimamoto (unpublished). The amplification factor, w/w_0, is plotted against the average strain of the system, ϵ_s, in Figure 5.16 for a fixed viscosity ratio of 30 and for the variation of the maximum initial limb dip, α_0. The result from Biot's theory is shown by the broken line, and it can be seen that his prediction agrees reasonably well with the finite element result when α is small. As in the case of single-layer folds, the fold grows slowly at first and then increasingly rapidly after a certain stage. The result for $\alpha_0 = 4°$ indicates, however, that the rate of fold growth starts to decline when the maximum limb dip, α_{max},

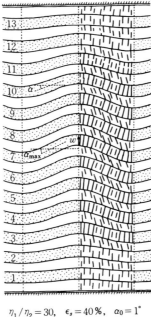

$$\eta_1 / \eta_2 = 30, \quad \epsilon_s = 40\%, \quad \alpha_0 = 1°$$

Figure 5.15   Finite-element analysis of strain pattern in intrafolial fold (after Shima-moto). The strain pattern is shown by orientations of the maximum principal elongation. Stippled areas: competent layers, η_1/η_2: viscosity ratio, ϵ_s: average strain of the system, α_{max}: maximum value of limb dip, α_0: initial α_{max}, w: maximum value of amplitude when the average strain of the system is ϵ_s

exceeds $30 - 40°$. This contradicts the results of Biot's linear theory and means that the fold cannot grow indefinitely.

The directions of the maximum elongation (i.e. the directions of the major axis, X, of the strain ellipsoid) throughout an intrafolial fold are shown by short lines in Figure 5.15. The strain distribution is characterized by the directions of X showing a fan-like arrangement which converges downward and upward on the anticlinal sections of competent and incompetent layers, respectively. Equation (5.19), derived from Biot's theory, gives only a strain distribution averaged over local heterogeneous strains due to the layering, and hence it is not very useful for comparisons with natural folds.

(d)   Intrafolial folds of Sambagawa crystalline schists

Let us finally examine the characteristics of natural intrafolial folds. Figure 5.17 shows the fold wavelength, $\mathcal{L}$, plotted against the confinement distance,

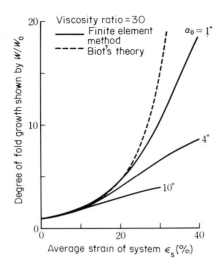

Figure 5.16 Growth processes of intrafolial folds obtained from Biot's theory (1965c*) and from numerical experiments based on finite-element analysis (after Shimamoto). Numerical experiments are performed on the model of Figure 5.15. α_0: initial value of maximum limb dip

H, for intrafolial folds in the interior of quartz schists. $\mathcal{L}$ and H fall within an area bounded by the broken lines in this diagram, and ξ, defined by equation (5.20), varies from a maximum of 0.44 (the upper line) to a minimum of 0.05 (the lower line). The parameter, ξ, clearly decreases as the folds grow (Figure 5.18).

The existence of a maximum for ξ can be explained by Biot's theory as follows. The elastic solution predicts that under the initial compressive stress, P_0, only folds with ξ of less than ξ_0 will form (Figure 5.13), and ξ_0 can be identified with the observed maximum of ξ. The viscous solution indicates that the rate of fold growth dramatically slows down when ξ exceeds 0.3 or so (Figure 5.14), so that folds with ξ greater than 0.44 could not develop in practice.

The meaning of the existence of a minimum for ξ will become clear when one examines how ξ changes as folds grow. Biot's theory of internal viscous buckling predicts that ξ decreases during the folding in accordance with equation (5.21). In order to compare his theory with the observed change in ξ, the relationship between the maximum interlimb angle, θ, and the parameter, ξ, was calculated by use of equations (5.22) and (5.23) for three viscosity ratios (Figure 5.18). The initial values of ξ and θ were assumed to be 0.45 and 135°, respectively, in the calculation. The observed change in ξ correlates roughly with the theoretical prediction for $\eta_1/\eta_2 = 1$. However, the

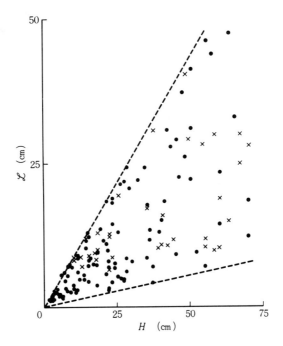

Figure 5.17   Relationship between $\mathcal{L}$ and H for intrafolial folds in siliceous schists of the Sambagawa belt (after Shimamoto). $\mathcal{L}$: wavelength, H: distance between non-folded layers by which the folded portion is confined. ●: folds with orthorhombic symmetry (Figure 5.12(a)), x: folds with monoclinic symmetry (Figure 5.12(b))

quartz schists are composed of alternated micaceous and quartzose layers whose viscosity ratio would be about 20 – 100 judging from the compositions of these layers and the results in Table 5.1. Even considering the strain distribution to be discussed below, in no way could the viscosity ratio be estimated as low as 1. Thus Biot's theory cannot account for the observed relationship between ξ and θ.

The principal source of this disparity is perhaps the assumption of the theory that AA′ and BB′ in Figure 5.12(a) remain non-folded during the folding. Excluding a special case in which the system is bounded above and below by very thick isotropic competent layers, this perhaps is not a valid assumption. It is more likely that the flat layers adjacent to the folded section become folded as the folding progresses, and if so, it is understandable that the observed ξ is much less than Biot's prediction. If in fact the folded section becomes widened during the folding in the direction normal to the layering, the confinement distance, H, increases and the parameter, ξ, decreases more rapidly than predicted by equation (5.21). Since the rate of fold growth is

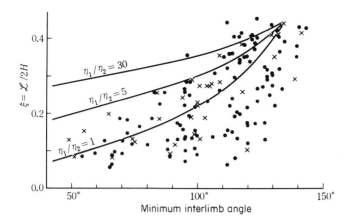

Figure 5.18 Relationship between $\xi = \mathcal{L}/2H$ and minimum interlimb angle θ for intrafolial folds in siliceous schist of the Sambagawa belt (after Shimamoto). •: folds with orthorhombic symmetry, x: folds with monoclinic symmetry. The solid lines show the relationship between ξ and θ for the internal buckling of the first kind in layers of equal thickness and alternate viscosity coefficients η_1 and η_2. This relationship is obtained from Biot's theory (1965c*)

larger for smaller ξ (Figure 5.14), the folding will be accelerated by this widening of the folded section. The folded section cannot be extended indefinitely because the neighbouring folded sections would also tend to expand. The minimum value for ξ is set presumably by this interaction between neighbouring folds.

In the folding of a single layer enclosed in an isotropic medium, the effects of the folding are transmitted through the medium for a distance no more than the fold wavelength (see Section 5.5). In marked contrast, many intrafolial folds have H up to ten times as large as $\mathcal{L}$ (Figure 5.17). Recall that the direction normal to the layering, i.e. the direction corresponding to $\xi = 0$, is the most unstable characteristic direction in the internal buckling of the first kind. Thus it may be possible to interpret the upward and downward expansion of the folded section as the propagation of local deformation related to the folding along this characteristic direction. However, no theory has yet been worked out to predict how the folded section expands during the intrafolial folding.

Another marked feature of intrafolial folds is that they have no dominant wavelength, unlike single-layer folds. However, because the resistance to bending of each competent layer is greater for smaller wavelengths, the fold wavelength cannot become infinitely small. As discussed in Section 5.8, huge folds with wavelengths exceeding several tens of kilometres do not form because of the stabilizing effect of gravity. Initial irregularity of layers are

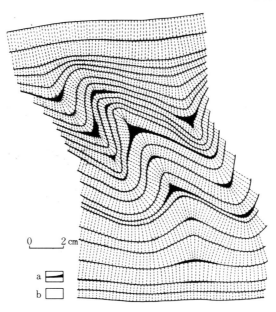

Figure 5.19   Strain pattern of intrafolial fold in siliceous schist of the Sambagawa belt (Hara, 1972*). a: incompetent mica-rich layer, b: competent quartz-rich layer, dotted lines: direction of maximum principal elongation

likely to be present on various scales, so that the folds would correspondingly possess various wavelengths excluding the above two extremes.

The patterns of schistosity or cleavage in intrafolial folds within quartz schists are basically the same as those seen in single-layer folds (cf. Figures 5.19 and 5.7) and agree reasonably well with the finite element result in Figure 5.15.

Thus the features of natural intrafolial folds can be explained fairly well by the mechanical model of Biot, and it may be concluded that the most important causative factor in the intrafolial folding is the mechanical anisotropy of the rocks as a whole.

5.4   Kink bands and conjugate folds

Two basically different types of folding have been recognized in multi-layers under layer-parallel compressive stress (Figure 5.20). One is the folding that produces folds with only a single axial surface, whereas folds with two sets of conjugate, often symmetrical axial surfaces form in the other type of folding. The former is known generally as folding, and the latter as conjugate folding or as kinking. For folds with rounded hinges '*conjugate folds*' would be the

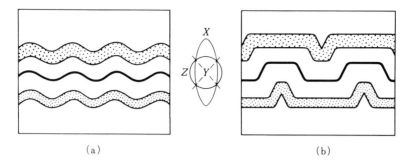

(a)   (b)

Figure 5.20   Two types of fold and their orientational relationship to the principal axes ($X>Y>Z$) of strain ellipsoid showing the average strain of the system. (a) folds with single axial surface, (b) conjugate folds (kink bands)

desirable term, but the sharply-bent folds with two sets of axial surfaces are generally called '*kink bands*'. The axial surfaces of kink bands are called kink planes. The section summarizes the characteristic features of conjugate folds and kink bands and discusses the mechanisms and physical conditions for their development.

(a)   Geometry and strain distribution

Kink bands may be classified into three types as shown in Figure 5.21 based on their geometrical features.

Type A kink bands (Figure 5.21 (a), (a′)) are typically seen in very fissile rocks (e.g. those produced experimentally by Paterson and Weiss, 1966*), and their characteristics are as listed below.

(1) Kink band grows from a minute embryonic one as shown from (a) to (a′) in Figure 5.21,
(2) Not only the width of kink band, S, but also its length, PQ, increases during the progressive deformation,
(3) The angles, γ and γ_k, defined in Figure 5.21 (b) remain almost unchanged throughout the deformation (in Paterson and Weiss' experiments $\gamma \simeq 58°$, $\gamma_k \simeq 65°$),
(4) The planar structures are very sharply bent along the kink planes. Intracrystalline kink bands also show similar features to them (McClintock and Argon, 1966*).

Type B kink bands (Figure 5.12 (b), (b′)) are typically found in slate and in crystalline schists with schistosity but without marked compositional banding (Donath, 1964, 1968), and their characteristics are:

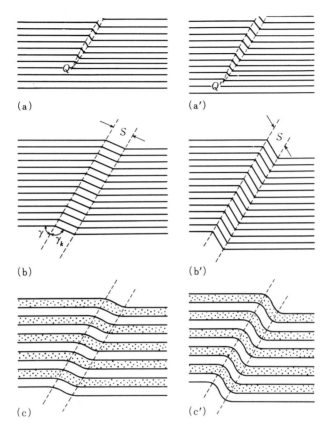

Figure 5.21   Forms of kink bands. The average strain of the system is larger in the right diagram than in the left diagram γ and γ_k: angles between kink plane and layer surface. (a) – (a'): kink bands found in crystals and highly fissile rocks. *PQ*: length of kink band. (b) – (b'): kink bands found in rocks which, although fissile, contains granular minerals in large amounts. *S*: width of kink band. (c) – (c'): kink bands found in layers of alternate mechanical properties

(1) The planar structures are sharply bent.
(2) γ_k does not remain constant but decreases during the formation (in Donath's experiments γ was held at almost 68° but γ_k decreased down to 55°)
(3) The width of the kink band, *S*, remains almost unchanged (*S* might just slightly increase) during the kinking.

The second and third points differ from the features of type A. Kink bands are also often formed in geologic bodies composed of two or more types of

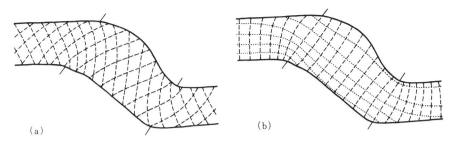

Figure 5.22   Analysis of strain pattern in kink band of calcite-rich layer (Hara and Paulitsch, 1971*). Strain is concentrated in domains along kink plane. (a) Orientation (dashed lines) of predominant set of $e(01\bar{1}2)$ lamellae in calcite grains. (b) Directions of maximum shortening (dotted lines) and maximum extension (dashed lines) obtained from e-lamellae

alternated layers. Type C kink bands (Figure 5.21 (c), (c′)) are typically found when alternating layers contain abundant granular minerals and lack fissility (Hara, 1965; Hara and Paulitsch, 1971*). Their characteristics are the same as those for type B kink bands except that these granular layers are not sharply bent along the kink planes. In the natural kink bands reported by Hara and Paulitsch, γ is almost constant at 60° but γ_k varies down to 29°. This type of kink bands are of considerable interest in that they show strain distributions similar to those of buckling folds (Figure 5.22). Type C kink bands are often called conjugate folds.

A very important feature common to all three types is that the angle between the kink planes and the direction of the maximum compressive stress is greater than 45°, unlike the case of faulting, and is commonly around 60°.

(b)   Mechanisms of kinking

The mechanisms of kinking are still not well understood. Studies of intracrystalline kinking have revealed that (1) the existence of a single slip plane in a direction prone to sliding and (2) the existence of some constraints which hinder the slippage along the slip plane, are both very important in the development of kink bands (see McClintock and Argon, 1966*). An illustrative example is shown in Figure 5.23 in which cylindrical specimens with a single slip plane are compressed between pistons. Kink bands will develop only in (b) and (c) where the slippage is restrained by the pistons. Paterson and Weiss (1966*) reached the same conclusion in their kinking experiments of phyllites. Hara and Paulitsch (1971*) studied kink bands in crystalline schists and found that the granular layers had been intensely deformed during the kinking (Figure 5.22), but they found no evidence for the occurrence of

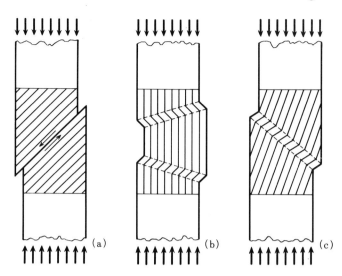

Figure 5.23   Deformation of highly fissile rocks. (a) When slippage along fissile planes is not confined by the piston. (b) When slippage along fissile planes is completely confined by the piston. (c) When slippage along fissile planes is partially confined by the piston

slippage along schistosity.† Thus, the slippage cannot always be regarded as necessary in the development of kink bands. The principal geometric features common to all kink bands are (1) that the direction of the kink plane, or the angle γ, remains almost unchanged during the kinking and (2) that the length of kink bands, PQ, is much larger than their width, S (Figure 5.21). This suggests that there is a common ground for the formation of all types of kink bands, even though they are variable in fine details. Indeed, it has been shown experimentally that at least type A and C kink bands form via a process in which a local deformation associated with embryonic kink bands propagates along a specific direction and that the above geometric features reflect such a process of kinking (Klassen-Neklyudova *et al.*, 1960*; Weiss, 1968; Reches and Johnson, 1976*; etc.). Recognizing that kink bands thus possess features of hyperbolic solutions explained in the previous section, Johnson (1970) and Cobbold *et al.* (1971) pointed out the possibility that kink bands form as a result of Biot's internal buckling of the second kind.

† Ductile deformation is predominant under metamorphic conditions. Slippage along discrete surfaces will not generally occur in such cases, because the frictional resistance is as high as the flow stress of the rock itself (see Orowan, 1960; Mogi, 1974).

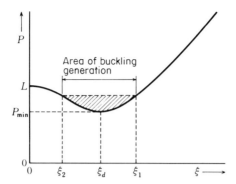

Figure 5.24 Relationship between ξ and external compressive stress P in generation of internal buckling of the second kind in anisotropic elastic body (Biot, 1965a[*])

The internal buckling of the second kind occurs when M and L defined by equation (5.15) satisfy (Biot, 1963[*], 1965a[*]):

$$L > 2M, \qquad L > P > 4M(L-M)/L \qquad (5.25)$$

In this case the compressive stress, P, that satisfies the characteristic equation (5.16) has a minimum at (Figure 5.24):

$$\xi_d = \{(L-2M)/L\}^{\frac{1}{2}} \qquad (5.26)$$

which gives the characteristic direction corresponding to the least compressive stress, P, i.e., the direction along which deformation is most prone to occur. From equations (5.25) and (5.26), $1 \geq \xi_d > 0$. Thus if this characteristic direction can be correlated with the direction of a kink plane, the theory predicts that the kink plane forms at an angle of $45 - 90°$ to the direction of P. The angle, γ, in Figure 5.21 is greater than $45°$ with almost no exceptions, so that Biot's theory accounts for the direction of kink planes at least qualitatively.

Quantitatively, however, the theoretical prediction is not very good. Application of Biot's theory to the experimental data of Paterson and Weiss (1966[*]) yields $\gamma = 46°$ ($\xi_d = 0.96$), whereas the observed value is $58°$. The major source for this discrepancy lies perhaps in the theoretical assumption that deformation is elastic just until the onset of kinking. In fact, Klassen-Neklyudova *et al.* (1960[*]) recognised, through careful observations, that plastic deformation precedes the onset of kinking. To analyse the mechanism of kinking, it would be necessary first to understand the nature of this plastic deformation and then to incorporate it with theoretical formulations.

Frank and Stroh (1952) proposed a theory of intracrystalline kinking based on the dislocation mechanics, but there has been no attempt to compare the theory with kink bands within rock-forming minerals.

(c)   The ambient conditions for kinking and folding

Let us finally consider the ambient physical conditions for the formation of kink bands. In the Sambagawa belt, as in many other regional metamorphic belts, kink bands formed in the final stage of deformation, following the formation of folds with single axial surfaces. Hara (1974) analysed deformation-lamella and c-axis fabrics of quartz in both kink bands and folds and found that deformation associated with the kinking took place at lower temperatures than those for the folding. This suggests that the mode of deformation changes from kinking to folding as the temperature rises, although there is a possibility that water pressure played a significant role. If the physical conditions required for the formation of kink bands and folds are established, it would be possible to deduce the general conditions prevailing at the time of kinking and folding. It is within our present technological capability to examine this kinking-folding transition, if fine-grained phyllites are used for specimens. It is thus hoped that the transition will be examined in the near future.

5.5   Folding of multilayered systems

Extensive researches have been conducted on the folding of multilayers. They are too voluminous to be reviewed here in detail, so only a brief account will follow.

Folding of multilayers due to the internal buckling has already been discussed in the previous two sections. Another simple example of multilayer-folds is harmonic folds of multiple competent layers enclosed in a thick incompetent medium. Biot (1965d*) called them '*similar folds*', although this usage of the term is somewhat different from that in structural geology (see Section 5.2(d)). He showed that when the folded multilayers consist of alternated competent and incompetent layers, the geometrical features of the folds depend on the viscosity ratio between the layers, and derived equations for the dominant wavelengths for large and small viscosity ratios. This type of folding is due to the buckling of the alternated multilayers as a whole, so that the folding mechanism resembles that of the single-layer folding. However, the nature of folding changes from buckling to internal buckling as the total thickness of the alternated multilayers becomes very much greater than that of the individual layers. Hara (1966*, 1967) and Roberts and Strömgård (1972*) analysed natural and model folds of this type and reported their stress and strain distributions.

What Biot called similar folds form when each incompetent layer between adjacent competent layers is sufficiently small. The greater the spacing of adjacent competent layers, the greater will be the tendency for individual competent layers to be folded independently. According to the theoretical and experimental analyses of Currie *et al.* (1962*) and Ramberg (1962*), two competent layers are folded independently of each other when their spacing is greater than the sum of the fold wavelengths of the two layers.† If the spacing of adjacent competent layers is less than this distance but is wider than that for the formation of harmonic folds, each competent layer tends to be folded independently while affecting the folding of adjacent competent layers, resulting in the formation of complex folds. Superposed folds of different scales are examples of these complex folds (Ramberg, 1964*, etc.), and such complex folds are those most commonly observed in nature. To analyse the mechanisms of complex folding, one must rely on a more general analysis than those discussed so far.

General theories of folding in multilayer systems have been worked out by Biot (1963, 1964, 1965a*), Ramberg (1970*), Johnson and Honea (1975), Johnson and Page (1976) and Johnson (1977*). These theories are based on the concept that the folding is mainly due to the differences in mechanical properties between layers. When interlayer slippage is thought to have been important, the theories have to be revised to incorporate the friction operating between layers. There are not many applications of these general theories at the moment and analysis of complex multilayer-folds in nature is left for the future. However, the work of Ramberg (1970*), who analysed multilayer-folds consisting of superposed folds of two different scales and that of Ramberg and Strömgård (1971*), who verified Ramberg's theoretical prediction experimentally are worthy of note.

5.6 Three-dimensional analysis of folding

Almost all studies of folds and folding have been concerned with the two-dimensional analyses in cross-sections perpendicular to the fold axis. Implicit in this work is that the nature of deformation does not change appreciably in the direction of the fold axis. In practice, however, the length of a fold axis is finite (Figure 5.25), and the geometry of a fold often changes in the axial direction. To be precise, therefore, analysis of folding should be undertaken in three dimensions. This section deals with the three-dimensional aspects of single-layer folding.

---

† This criterion applies only if the incompetent medium is isotropic. If the medium is anisotropic and if the internal buckling can occur within it, heterogeneous deformation associated with the folding can be propagated in the medium over a distance much greater than the fold wavelength.

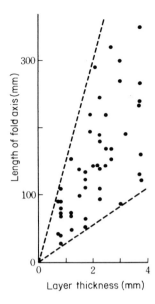

Figure 5.25 Relationship between length of fold axis and layer thickness for single-layer folds. Data from folds of quartz veins enclosed in psammitic schist of the Sambagawa belt (Hara *et al.*, 1975*)

Let us begin with reviewing the analytical result of Ghosh (1970*). He considered the folding of a viscous competent layer enclosed in a thick viscous medium under lateral compressive stresses, P_x and P_y, acting in the x and y directions, respectively, where x and y co-ordinate axes are taken parallel to the layer. The effects of gravity and uniform shortening of the layer are ignored. Ghosh also assumed that the layer would undergo sinusoidal deflections in both x and y directions, and dealt with the initial deflection, w_0, given by

$$w_0\ (x,y) = A_0 \sin l_1 x \sin l_2 y. \tag{5.27}$$

This initial deflection grows exponentially with time, i.e.

$$w = w_0 e^{pt} = A_0 e^{pt} \sin l_1 x \sin l_2 y \tag{5.28}$$

where the wavenumbers, l_1 and l_2, are defined in terms of the fold wavelengths, $\mathscr{L}_x$ and $\mathscr{L}_y$, in the x and y directions as

$$l_1 = 2\pi/\mathscr{L}_x, \qquad l_2 = 2\pi/\mathscr{L}_y \tag{5.29}$$

In deriving equation (5.28), it is assumed that $\mathcal{L}_x$, and $\mathcal{L}_y$ do not change during the folding process. Equation (5.28) is of the same form as the two-dimensional solution, equation (5.1), and p is given by

$$p = \frac{P_x \, (s^2 + Bq^2)}{\frac{1}{3}\eta_1 \, (s^2 + q^2)^2 + 4\eta_2 \, \sqrt{s^2 + q^2}} \tag{5.30}$$

with

$$s \quad = l_1 h = 2\pi h/\mathcal{L}_x$$

$$q \quad = l_2 h = 2\pi h/\mathcal{L}_y \tag{5.31}$$

$$B = P_y/P_x \tag{5.32}$$

The dominant wavelength should maximize p, as in the two-dimensional analysis. Consider first the case when $B = 1$, i.e. when the layer is subjected to equal compressive stresses in the x and y directions ($P_x = P_y$). Assuming $\mathcal{L}_x = \mathcal{L}_y$ based on the symmetry consideration, the dominant wavelength is expressed as

$$\mathcal{L}_x = \mathcal{L}_y = 2\sqrt{2}\pi h \sqrt[3]{\frac{\eta_1}{6\eta_2}} = \sqrt{2}\mathcal{L}_d \tag{5.33}$$

where $\mathcal{L}_d$ is the dominant wavelength obtained from the two-dimensional theory and is given by equation (5.3).

Next consider the more general case when $1 > B \geq 0$. The dominant wavelengths in the x and y directions can be derived from the stationary conditions for p ($\partial p/\partial s = 0$, $\partial p/\partial q = 0$) are given by:

$$\mathcal{L}_x = 2\pi h \sqrt[3]{\frac{\eta_1}{6\eta_2}} = \mathcal{L}_d, \qquad \mathcal{L}_y = \infty \tag{5.34}$$

Even under the compressive stress, P_y, in the direction of fold axis, the rate of fold growth is maximum when the fold axis is infinite. Also note $\mathcal{L}_x = \mathcal{L}_d$ in this case.

In order for a fold to possess an infinite fold axis, $\mathcal{L}_y$ has to be infinite already at the initial undeformed state. This is very unlikely, however, so the fold axis will inevitably be of finite length. Based on this physical consideration, let us assume that $\mathcal{L}_y$ is finite. Then the dominant wavelength, $\mathcal{L}_x$, for a fixed $\mathcal{L}_x/\mathcal{L}_y$ is given by (Ghosh, 1970*)

$$\mathcal{L}_x/\mathcal{L}_d = \sqrt{1 + (\mathcal{L}_x/\mathcal{L}_y)^2} \tag{5.35}$$

or

$$(1/\mathcal{L}_x)^2 + (1/\mathcal{L}_y)^2 = (1/\mathcal{L}_d)^2 \qquad (5.36)$$

which reduces to equation (5.33) when $\mathcal{L}_x = \mathcal{L}_y$ and to equation (5.34) when $\mathcal{L}_y = \infty$ (Figure 5.26). These analytical results of Ghosh (1970*) have two important bearings on folding in three dimensions. First, a fold grows at the maximum rate when its fold axis is infinite ($\mathcal{L} = \infty$). Ghosh implicitly assumed that a certain fixed value of $\mathcal{L}_x/\mathcal{L}_y$ corresponds to a constant value of $B = P_y/P_x$. It is more likely, however, that the fold axis becomes longer during the folding, for the longer the axis the more rapid will be the growth of the fold. Indeed, Hara *et al.* (1975*) suggest, on the basis of their observations of natural folds, that the lengths of the fold axes increase with the growth of the folds. The second point is whether or not the two-dimensional theory can be applied to the folds with fold axes of finite lengths. According to Hara *et al.* (1975*), $\mathcal{L}_x/\mathcal{L}_y$ for folds of quartz veins in psammitic schists is $0.1 - 0.3$ when $\mathcal{L}_d/h$ is $8 - 8.5$ and the limb dip is about $30°$. Thus, from equation (5.35), $\mathcal{L}_x/\mathcal{L}_d = 1.005 - 1.044$, and the dominant wavelength, $\mathcal{L}_x$, predicted from the three-dimensional theory differ by no more than a few per cent from the dominant wavelength, $\mathcal{L}_d$, given by the two-dimensional theory. The process of wavelength selection and the change in the length of fold axis during the folding have not yet been examined in sufficient detail in three-dimensions, therefore, strictly speaking the validity of the two-dimensional theory cannot yet be conclusively evaluated. However, Ghosh's theoretical analysis combined with the observations of Hara *et al.*, suggests

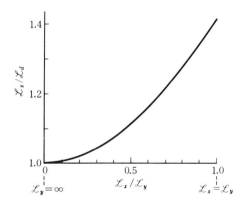

Figure 5.26 Relationship between three-dimensional solution and two-dimensional solution of single-layer folding based on Ghosh (1970*). $\mathcal{L}_y$: length of fold axis, $\mathcal{L}_x$: wavelength, $\mathcal{L}_d$: dominant wavelength of two-dimensional solution

that the two-dimensional theory can be applied to natural folds with reasonable accuracy at least in regard to the dominant wavelength. Three-dimensional analyses have just begun not only in the analyses of natural folds but also in the theoretical and experimental approaches. Three-dimensional treatise of folding in more complex systems must await future studies.

5.7 Bending folds and drape folds

We have been concerned so far only with various types of buckling folds produced by lateral compression. This section describes another type of fold, i.e. bending folds produced by heterogeneous compressive stresses acting perpendicularly or nearly perpendicularly to the layered structures. Bending folds are found in the strata pierced by salt domes, in the strata overlying faulted basement blocks, or in the gneiss layers around mantled gneiss domes. Examples of bending folds associated with salt domes in Texas are shown in Figure 5.27. The structural features of bending folds have been studied theoretically, and experimentally by many workers since the publication of Hafner's (1951*) theory (Parker and McDowell, 1955; Sanford, 1959*;

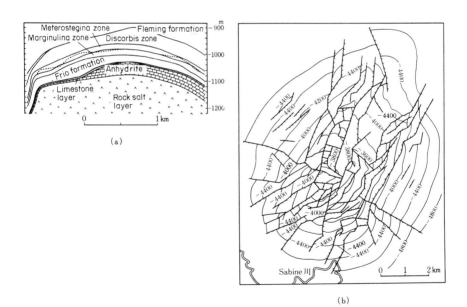

Figure 5.27   Bending fold of covering layers produced by the rise of salt dome. (a) profile of bending fold seen with salt dome in Sugar¹and oilfield, Texas (McCarter and O'Bannon, 1933). (b) radial faults in bending fold over Hawkins salt dome (Went-landt, 1951). Fold form is shown by depth contours (in feet) of the top of salt layer

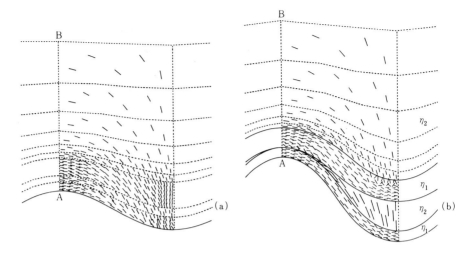

Figure 5.28   Strain analysis of viscous bending folds by finite-element method (Ikeda and Shimamoto, 1974*). Strain patterns are shown by orientation of maximum principal elongation. (a) Bending fold in uniform system. Average shortening between AB = 10%. (b) Bending fold in system containing two competent layers. Viscosity ratio between competent layers and medium η_1/η_2 = 100. Average shortening between AB = 20%

Ramberg, 1963*; Fletcher, 1972; Ikeda and Shimamoto, 1974*; Couples, 1977).

Figure 5.28 shows results from the finite element analysis of the bending folds produced in a layered media composed of Newtonian fluids due to sinusoidal displacement from below. The shape of the underlying geologic body is directly reflected in the geometry of the folds. The folds amplitude diminishes upward, and the folds disappear near the top boundaries. As the displacement on top of the underlying geologic body proceeds, the horizon at which the fold disappears moves upward. Strain distributions in bending folds are characterized by the gradual change in the direction of maximum elongation (schistosity or cleavage is expected to form in this direction at high strains if the layers are ductile) from a direction parallel to the layers in the anticlinal zone to a direction perpendicular to the layers in the synclinal zone (Figure 5.28(a)). The layers thus become thinner in the anticlinal zone and thicker in the synclinal zone reflecting this strain distribution. Competent layers are almost uniformly elongated resulting in no marked variation in the thickness distributions, whereas incompetent layers between them show marked increase and decrease in layer thickness in the synclinal and anticlinal zones, respectively (Figure 5.28(b)). The direction of maximum elongation within the competent layers is nearly parallel to the layer in most places. The

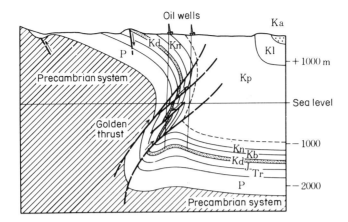

Figure 5.29   Drape fold in covering layers produced by thrusting in underlying layers. This example is related to Golded thrust near Soda Lakes, Colorado (from Berg, 1962). P: Palaeozoic formation, Tr, J, Kd–Ka: Mesozoic formation

strain distribution in Figure 5.28(a) resembles the stress distribution in elastic bending folds analysed by Hafner (1951*). Based on this stress distribution, Hafner pointed out that normal and reverse faults would form in anticlinal and synclinal zones, respectively. The normal radial faults around the dome shown in Figure 5.27(b) can presumably be interpreted similarly.

Stephansson (1971) examined bending folds of limestone layers associated with mudstone diapirs on Oland Island in southern Sweden, and reported fold geometries and thickness distributions very similar to those shown in Figure 5.28. He recognized that when a system consists of alternated limestone and mudstone layers, the limestone behaved as competent layers and the mudstone as incompetent layers. Changes in their layer thicknesses were also strikingly similar to the results in Figure 5.28(b). Unfortunately, there are no good examples of natural bending folds, to the authors' knowledge, which have been analysed in sufficient detail to permit a comparison of their stress and strain distributions with theoretical and experimental results.

A bending fold of the overlying strata due to the differential movements of faulted basement blocks is often called a '*drape fold*', for the folded strata cover the uplifted basement blocks like a tablecloth. Observe in an example of natural drape fold shown in Figure 5.29 that the drape fold developed concordantly with the shape of the uplifted block. Many drape folds have been reported in the western United States (Prucha *et al.*, 1965; Stearns, 1971; Stearns and Weinberg, 1975; etc.). Tsuneishi (1966) reports similar geological structures on the eastern edge of the Abukuma Plateau, Japan.

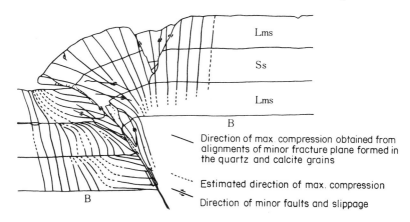

Figure 5.30 Forms and stress patterns of drape folds produced by triaxial experiments (Friedman *et al.*, 1976*). The diagram shows drape folding and subsequent faulting in two layers of limestone (Lms) and a sandstone layer (Ss) induced by thrusting of basement rocks (B)

These drape folds formed at shallow depths, so that the deformation of both the basement and the overlying strata is predominantly brittle. Even though the strata are folded as a whole, numerous faults develop very frequently within the strata. The development of faults is particularly marked in the areas adjacent to the basement faults (Figure 5.29).

Deformation of the overlying strata due to the differential movements of basement blocks has been investigated experimentally by Gzovsky and Ma-Chin (1968), Kodama *et al.* (1974, 1976) and Friedman *et al.* (1976*). Friedman *et al.* produced structures (Figure 5.30) very similar to natural drape folds (Figure 5.29) with respect to the geometry of the folded overlying strata and to the mode of fault development. In the experimental drape fold in Figure 5.30, a graben formed in the overlying layers above the uplifted block reflecting tensile or extensile stress parallel to the layers there. A structure remarkably similar to the graben in Figure 5.30 has been found in a drape fold in the Rattlesnake Mountain in Wyoming. Friedman *et al.* analysed stress distribution in the drape fold from the directions of microfractures developed in quartz and calcite grains (Figure 5.30). Gangi *et al.* (1977) also obtained stress distribution theoretically, and is in reasonable agreement with the experimental results. Unfortunately, however, no detailed analyses of stress distribution in natural drape folds have yet been published (possibly due to the threatening of rattlesnakes).

It should be noted that the bending folds and drape folds are markedly different from buckling fold in regard to their geometry and stress and strain distributions. Because difference in the mechanism of folding is reflected in such structural characteristics, it is possible to infer the mechanisms for natural folds by examining their structural characteristics. Acidic magma emplaced concordantly in layered rocks often takes on a form of a low dome with flat base because of its high viscosity, and is known as '*laccolith*' (e.g. Billings, 1972). The strata just above the magma are pushed up to form a dome-shaped structure, which is a type of bending fold. The mechanical processes involved in the formation of laccoliths have been examined in detail by Johnson (1970), Johnson and Pollard (1973*) and Pollard (1973).

5.8 Effects of gravity on buckle folding

All materials on and in the earth are under the action of the earth's gravity, whose effects must therefore be taken into account in any analysis of folding. Smoluchowski (1909, 1910), Biot (1961*), Ramberg and Stephansson (1964) and Ramberg (1967*, 1970*) studied theoretically and experimentally the effects of gravity on both single-layer and multilayer folding. This section is intended to elucidate the essential features of the effects of gravity by going through Biot's (1961*) analysis on the single-layer folding as a simple illustrative example.

Consider the folding of a system consisting of a single layer and underlying incompetent medium under the action of lateral compression and gravity (Figure 5.31). The reason for not including the medium on top of the layer is that the analysis is intended to simulate folding near the earth's surface. Assuming a sinusoidal fold shape, the parameter, p, which indicates the rate of fold growth is given by (Biot, 1961*):

$$p = \left(P - \frac{\rho_2 g}{h l^2}\right) \bigg/ \left(\frac{2\eta_2}{hl} + \frac{1}{3}\eta_1 h^2 l^2\right) \tag{5.37}$$

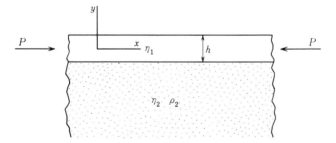

Figure 5.31 System comprising a single layer subjected to compressive stress P and an underlying medium. η_1, η_2: viscosity coefficients of layer and medium, ρ_2: specific gravity of medium, h: layer thickness, x, y: co-ordinate system

The difference between this equation and equation (5.2) is the inclusion of the term, $\rho_2 g/hl^2$, related to the effect of gravity on the right-hand side. This term always lower p; that is, gravity tends to hinder the growth of a fold (stabilizing effect of gravity). The medium on top of the layer is missing in the present model, and hence the first term in the denominator in equation (5.37) is half of that in equation (5.2). The dominant wavelength maximizes p, and one obtains from equation (5.37):

$$\left(\frac{P}{\rho_2 gh}\right)(2\pi)^5 \left(\frac{h}{\mathcal{L}_d}\right)^5 - 2(2\pi)^3 \left(\frac{h}{\mathcal{L}_d}\right)^3 -$$

$$3\left(\frac{\eta_2}{\eta_1}\right)\left(\frac{P}{\rho_2 gh}\right)(2\pi)^2 \left(\frac{h}{\mathcal{L}_d}\right)^2 - 3\left(\frac{\eta_2}{\eta_1}\right) = 0 \qquad (5.38)$$

The dominant wavelength–thickness ratio, $\mathcal{L}_d/h$, is given as the solution to this fifth-order algebraic equation that contains the viscosity ratio, η_1/η_2, and a non-dimensional product, $P/\rho_2 gh$, indicating the effect of gravity. The equation is solved numerically for $\mathcal{L}_d/h$ assuming various values of η_1/η_2 and $P/\rho_2 gh$, and the results are shown in Figure 5.32.

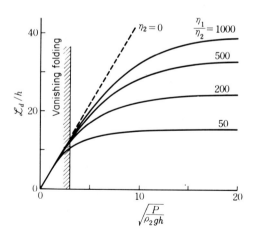

Figure 5.32   Dominant wavelengths $\mathcal{L}_d$ of single-layer folds affected by gravity (Biot, 1961*). *P*: compressive stress parallel to layer, η_1/η_2: viscosity ratio between layer and medium, ρ_2: specific gravity of medium underlying competent layer

Special cases will be considered below. Neglecting gravity in equation (5.38) leads to

$$\frac{\mathcal{L}_d}{h} = 2\pi \sqrt[3]{\frac{\eta_1}{3\eta_2}} \qquad (5.39)$$

which differs from equation (5.3) only in the integer factor in the cube root due to the absence of the overlying medium in this case. When the viscosity coefficient of the medium, η_2, is negligibly small relative to that of the layer, η_1, equation (5.38) reduces to:

$$\frac{\mathcal{L}_d}{h} = \pi \sqrt{\frac{2P}{\rho_2 gh}} \qquad (5.40)$$

This equation has been verified experimentally by Ramberg and Stephansson (1964). Equation (5.40) is shown as the broken line in Figure 5.32. Let us now examine the effects of gravity by inspecting Figure 5.32. First, it can be seen that not only the viscosity ratio but also the lateral compressive stress, P, affects the dominant wavelength. The effect of gravity enters through the parameter, $P/\rho_2 gh$, containing the layer thickness, h. Hence h is another controlling parameter in determining the significance of the gravity effect; in fact, the scale dependence of the gravity effect is governed by this quantity. As an example, consider the case when $\eta_1/\eta_2 = 50 - 200$. If $(P/\rho_2 gh)^{1/2}$ is greater than 20, equations (5.38) and (5.39) give almost the same values of $\mathcal{L}_d/h$ and therefore the effect of gravity is not important. If $P = 1$ kb, $\rho_2 = 2.5$ g/cm^3 and $g = 980$ cm/s^2, the above condition is equivalent to $h < 10$ m or $\mathcal{L}_d < 160 - 260$ m. Evidently, the effect of gravity diminishes as the size of a fold becomes smaller and smaller.

As the thickness of the layer, h, increases, $P/\rho_2 gh$ decreases to lower $\mathcal{L}_d/h$ (Figure 5.32). When $P/\rho_2 gh$ is close to 9, $\mathcal{L}_d/h$ becomes little affected by the viscosity ratio and approaches a value of about 10. Based on equation (5.37), Biot (1961[*]) demonstrated that the greater the size of a fold, the more gravity would hinder the growth of a fold, and that no obvious folds would develop when $P/\rho_2 gh < 9$.

Ramberg (1970[*]) investigated the effect of gravity on multilayer folding as well as on single-layer folding, and concluded that the effect is negligible when the fold wavelength is less than 100 m or so and that folds with wavelengths in excess of 30 km or so would not form owing to the stabilizing effect of gravity.

The results of Biot and Ramberg are very significant in structural geology at least for the following two reasons. First, folding models ignoring the effects of gravity can be applied to small-scale folds. Secondly, the stablilizing effect of gravity set an upper limit of 30 km or so for the wavelengths of buckling folds developed under lateral compression. Let us examine the second point in more detail. Kaizuka (1967*) studied large-scale Quaternary folds with wavelengths of 0.7 – 1100 km in Japan, and found that the larger the wavelength, the slower the growth rate. Hara *et al.* (1977) recognized that folds formed in the Sambagawa belt during the Hijikawa phase of deformation are predominantly small in scale, their wavelengths being mostly less than 8 km, and that folds with wavelengths exceeding 20 km were non-existent. Although it is unclear whether all of the folds examined by Kaizuka are buckle folds, those examined by Hara *et al.* are presumably buckle folds judging from the strain distribution in them. The fact that folds with wavelengths exceeding 20 km are not found in the Sambagawa belt, combined with the theoretical prediction that the larger the fold the slower will be its growth rate, suggests that gravitational stabilization is at work in natural folding. Ramberg and Stephansson (1964) and Ramberg (1967*) point out that huge 'geosynclines' are most unlikely to be formed by buckling of the crust under the action of horizontal compression.

5.9   Diapir folds

Just as material heavier than water cannot float on the water surface, so geological systems composed of low density layers and overlying layers of high density are unstable in the gravitational field. The low density materials tend to rise differentially under their own buoyancy and so the layer surface takes on remarkable wavy forms. These wave-like structures are folds and they are called diapir folds. Typical examples include salt domes, mud domes and gneiss domes. This section is intended to illustrate the characteristics of diapir folds by taking salt domes as an example (Figure 5.33).

Trusheim (1957, 1960*) studied salt domes in northern Germany, and showed that overburdens about 1000 m thick seemed to be necessary to produce the development of diapir folds in salt layers around 300 m thick. This would suggest that a salt layer begins to rise at the stage when its overlying sedimentary material having been buried and compacted achieves a density and a thickness which imparts buoyancy to the salt layer in magnitude enough to rise. According to Trusheim's observations, the rise of the salt layer appears to begin with the formation of salt pillows due to horizontal migration of the salt. The profile of the upper surface of the salt layer then appears to develop into the formation of columnar structures, i.e. salt plugs (stocks), and then of salt walls which were produced by connecting these together (Figure 5.34). One particular feature of such salt domes (salt pillows, plugs or walls) is that they have a regular distribution. In other words, the waveform of the upper surface of the salt layer produced during the early stages of the dome formation exhibits what may be described as a dominant

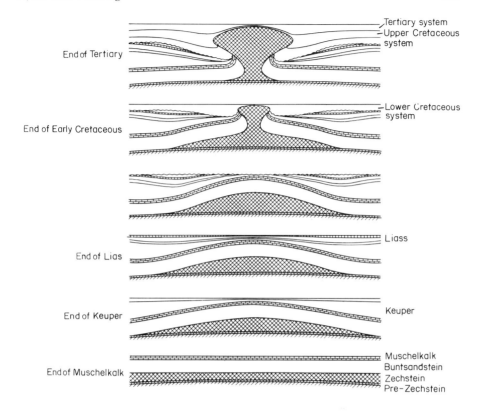

Figure 5.33 Schematic diagram showing development of salt dome in northern Germany (Trusheim, 1960*)

wavelength. The salt domes in the Mexican Gulf Coast have a dominant wavelength of about 10 km (Fletcher, 1972).

The development of salt domes leads to the formation of anticlines and synclines in the overlying strata. The magnitude of sedimentation differs markedly on the anticlines from in the synclines around the domes. The difference in thickness between the sediments on the domes and of the peripheral basins appears to be of order of 1 : 10. Assuming that the volume of sediments of the peripheral basins would equal the volume of salt migrating into the dome, Sannemann (1963) calculated the average rate of migration of salt to be 0.3 mm/year. The average salt migration rate obtained in such the way seems to be constant both during the salt pillow stage and during the salt plug stage. The value calculated by Sannemann is only about 1/10 that

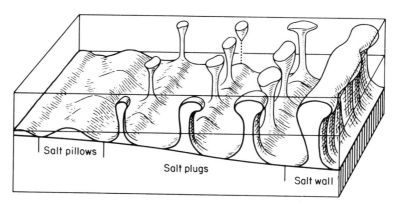

Figure 5.34   Schematic diagram showing relationship between burial depths of salt
layers and the forms of dome development (Trusheim, 1960*)

obtained by Lees and Falcon (1952), Lotze (1957) and Teichmuller *et al.*
(1958) for the current rise of salt plugs in many places. It has been clarified
that the salt structures in northern Germany are closely related to the depth
of burial of the salt layer and the form of the sedimentary basin (Figure 5.34)
(Trusheim, 1960*). Where the salt layer is buried to a depth in excess of 5000
m, it appears to form salt walls extending for more than 100 km in length.
They are 4–5 km wide and spaced at regular intervals of 8–10 km. The strike
of the salt walls is approximately parallel to the depth contours of the base of
the salt layer. The area in which the salt walls develop is surrounded by an
area of salt plugs. In this area, the base of the salt layer has been estimated to
be 3000–5000 m deep. The salt plugs are 2–8 km in diameter and tend to be
arranged in straight lines running parallel to each other. This is a phenomena
similar to the almost equally spaced parallel arrangement of the salt walls.
Outside the area where the salt plugs are developing, one only sees the
development of the salt pillows. Here the base of the salt layer is much
shallower. One may thus conclude that the form of salt domes is closely
related to the extent to which the salt had been buried.

The rise of low-density strata in the gravitational field has been studied
experimentally by Nettleton (1934, 1955), Debrin (1941), Parker and
McDowell (1955), Ramberg (1967*), Stephansson (1972), Berner *et al.* (1972)
Fletcher (1972), Talbot (1974, 1977) and Hayashi *et al.* (1975), and theoretically by Danes (1964), Biot and Odé (1965*), Biot (1966*) and Ramberg
(1967*, 1968). This section briefly reviews the two-dimensional analysis of
Biot and Odé (1965*) in order to understand the fundamental and general
features of diapir folding.

Biot and Odé analysed the behaviour of a system composed of a rigid
basement overlain by two viscous layers. The upper and lower layers have,
respectively, the thicknesses, h_1 and h_2, the viscosity coefficients η_1 and η_2,

and the densities, ρ_1 and ρ_2. The gravitational acceleration is assumed to be constant throughout the system. If $\rho_1 > \rho_2$, the underlying layer would rise in a waveform pushing aside the upper layer. Theory deals only with the initial stage of this rise. The upper surface of the low-density layer would normally have slight, irregular undulations in the initial state. The growth rate of these initial undulations depends on their wavelengths, and the wavelength corresponding to the maximum rate of growth (= dominant wavelength) will grow selectively as the time goes by. This dominant wavelength determines the spacing of the salt domes. Theory has shown that growth of the wavy undulations at first proceeds slowly but then accelerates. The dominant wavelength $\mathscr{L}_d$ is mainly determined by the viscosity ratio, η_1/η_2, and the layer-thickness ratio, h_1/h_2, that is, the larger η_1/η_2 and h_1/h_2, the larger will be $\mathscr{L}_d/h_2$. The parameter, $\Delta\rho/\rho_1 = (\rho_1 - \rho_2)/\rho_1$, which is an indicator of the density contrast, has almost no effect on the dominant wavelength, but the larger this parameter the higher will be the growth rate. Furthermore, the smaller η_1/η_2 or the greater h_1/h_2, the higher will be the growth rate. As mentioned above, Trushiem (1960*) recognized that sedimentation occurs in the peripheral basins, reflecting the underground movement. Biot and Odé extended the theory to take into account these surface phenomena, and showed that such phenomena bring about an increase in the growth rate and a very slight increase in the dominant wavelength.

Biot (1966*) made an ingenious use of the symmetry of the three-dimensional dome structures and the principle of superposition for the solutions of linear differential equations to extend the two-dimensional analysis of Biot and Odé into a three-dimensional treatment. His results demonstrate that the two-dimensional analysis is approximately valid at least in regard to the growth rate of domes and the distance between adjacent domes (or dominant wavelength).

The theory of Biot and Odé (1965*) predicts some features of natural salt domes quantitatively, especially in regard to the condition for the onset of diapir folding and the existence of the dominant wavelength (see their paper for further discussions). However, theories put forward so far cannot explain the origin of the regular zonal arrangements of salt walls and plugs such as those in Figure 5.34. In order to explain not just the rising processes of individual salt domes but their total evolutional processes in a certain area, one has to comprehend the whole tectonic framework for the complete sedimentary basins and then to incorporate it with mechanical model. Finally, it should be pointed out that the motion of granitic magmas presumably can be treated as the rising of low-density materials in a gravitational field (Ramberg, 1967*). This is another area of research which awaits further studies.

Ikuo Hara and Toshihiko Shimamoto

References

Biot, M. A. (1961): Theory of folding of stratified viscoelastic media and its implications in tectonics and orogenesis, *Geol. Soc. Am. Bull.*, **72**, 1595–1620.

Biot, M. A. (1963): Internal buckling under initial stress in finite elasticity, *Proc. Royal Soc.*, A 273, 306–28.

Biot, M. A. (1965a): *Mechanics of Incremental Deformations*, 504 pp., John Wiley, New York.

Biot, M. A. (1965b): Theory of viscous buckling and gravity instability of multilayers with large deformation, *Geol. Soc. Am. Bull.*, **76**, 371–8.

Biot, M. A. (1965c): Internal instability of anisotropic viscous and visco elastic media under initial stress, *J. Franklin Inst.*, **279**, 65–82.

Biot, M. A. (1965d): Theory of similar folding of the first kind and second kind, *Geol. Soc. Am. Bull.*, **76**, 251–8.

Biot, M. A. (1966): Three-dimensional gravity instability derived from two-dimensional solutions, *Geophys.*, **31**, 153–66.

Biot, M. A. and Odé, H. (1965): Theory of gravity instability with variable overburden and compaction, *Geophys.*, **30**, 213–27.

Currie, J. B., Patnode, H. W. and Trümp, R. P. (1962): Development of folds in sedimentary strata, *Geol. Soc. Am. Bull.*, **73**, 655–74.

Dieterich, J. H. (1969): Origin of cleavage in folded rocks, *Am. J. Sci.*, **267**, 129–54.

Friedman, M., Handin, J. W., Logan, J. M., Min, K. D. and Stearns, D. W. (1976): Experimental folding of rocks under confining pressure: Part III. Faulted drape folds in multilithologic layered specimens, *Geol. Soc. Am. Bull.*, **87**, 1049–66.

Ghosh, S. K. (1970): A theoretical study of intersecting fold patterns, *Tectonophys.*, **9**, 559–69.

Hafner, W. (1951): Stress distributions and faulting, *Geol. Soc. Am. Bull.*, **62**, 373–98.

Hara, I. (1966): Movement picture in confined incompetent layers in flexural folding — Deformation of heterogeneously layered rocks in flexural folding(I), *J. Geol. Soc. Japan*, **72**, 363–9.

Hara, I. (1972): Strain distribution in intrafolial folds, *J. Geol. Soc. Japan*, **78**, 531–9.

Hara, I., Ikeda, Y., Kimura, T. and Takeda, K. (1973): A note on folding style of multilayered rocks, *J. Geol. Soc. Japan*, **79**, 727–33.

Hara, I. and Paulitsch, P. (1971): Strain distribution in conjugate kink-bands, *N. Jb. Geol. Palaont. Abh.*, **139**, 346–8.

Hara, I., Uchibayashi, S., Yokota, Y., Umemura, H. and Oda, M. (1968): Geometry and internal structures of flexural folds; (I) Folding of a single competent layer enclosed in thick incompetent layer, *J. Sci. Hiroshima Univ., C.*, **6**, 51–113.

Hara, I., Yokoyama, S., Tsukuda, E. and Shiota, T. (1975): Three-dimensional size analysis of folds of quartz veins in the psammitic schist of the Oboke district, Shikoku, *J. Sci. Hiroshima Univ, C*, **7**, 125–32.

Hudleston, P. J. (1973): An analysis of "single-layer" folds developed experimentally in viscous media, *Tectonophys.*, **16**, 189–214.

Hudleston, P. J. and Stephansson, O. (1973): Layer shortening and fold-shape development in the buckling of single layers, *Tectonophys.*, **17**, 299–321.

Ikeda, Y. and Shimamoto, T. (1974): Numerical experiments on viscous bending folds, *Jour. Geol. Soc. Japan*, **80**, 65–74.

Johnson, A. M. (1977): *Styles of Folding*, 406 pp., Elsevier, Amsterdam.

Johnson, A. M. and Pollard, D. D. (1973): Mechanics of growth of some laccolithic intrusions in the Henry Mountains, Utah, *Tectonophys.*, **18**, 216–309.

Kaizuka, S. (1967): Rate of folding in the Quaternary and present, *Geogr. Rept. Tokyo Metrop. Univ.*, **2**, 1–10.

Klassen-Neklyudova, M. V., Chernysheva, M. A. and Tomilovskii, G. E. (1960): The mechanism of kinking, *Soviet Phys. Cryst. (English Transl.)*, **5**, 617–21.

McClintock, F. A. and Argon, A. S. (1966): *Mechanical Behavior of Materials*, 770 pp., Addison-Wesley, Canada.

Paterson, M. S. and Weiss, L. E. (1966): Experimental deformation and folding in phyllite, *Geol. Soc. Am. Bull.*, **77**, 343–74.

Parrish, D. K., Krivz, A. L. and Carter, N. L. (1976): Finite element folds of similar geometry, *Tectonophys.*, **32**, 183–207.

Ramberg, H. (1962): Contact strain and folding instability of a multilayered body under compression, *Geol. Rundschau*, **51**, 405–39.

Ramberg, H. (1963): Strain distribution and geometry of folds, *Bull. Geol. Inst. Univ. Uppsala*, **42**, 1–20.

Ramberg, H. (1964): Selective buckling of composite layers with contrasted rheological properties; a theory for simultaneous formation of several orders of folds, *Tectonophys.*, **1**, 307–41.

Ramberg, H. (1967): *Gravity, Deformation and the Earth's Crust*, 214 pp. Academic Press, London.

Ramberg, H. (1970): Folding of laterally compressed multilayers in the field of gravity, I, *Phys. Earth Planet. Int.*, **2**, 203–32.

Ramberg. H. and Strömgård, K. E. (1971): Experimental tests of modern buckling theory applied on multilayered media, *Tectonophys.*, **11**, 461–72.

Ramsay, J. G. (1967): *Folding and Fracturing of Rocks*, 568 pp., McGraw-Hill, New York.

Reches, Z. and Johnson, A. M. (1976): A theory of concentric, kink and sinusoidal folding and of monoclinal flexuring of compressible, elastic multilayers, *Tectonophys.*, **35**, 295–334.

Roberts, D. (1971): Abnormal cleavage patterns in fold hinge zones from Varanger Peninsula, Northern Norway, *Am. J. Sci.*, **271**, 170–80.

Roberts, D. and Strömgård, K. E. (1972): A comparison of natural and experimental strain patterns around fold hinge zones, *Tectonophys.*, **14**, 105–20.

Sanford, A. R. (1959): Analytical and experimental study of simple geologic structures, *Geol. Soc. Am. Bull.*, **70**, 19–52.

Schapery, R. A. (1964): Application of thermodynamics to thermomechanical, fracture, and birefringent phenomena in viscoelastic media, *J. Appl. Phys.*, **35**, 1451–65.

Shimamoto, T. and Hara, I. (1976): Geometry and strain distribution of single-layer folds, *Tectonophys.*, **30**, 1–34.

Trusheim, F. (1960): Mechanism of salt migration in Northern Germany, *Am. Assoc. Petrol. Geol. Bull.*, **44**, 1519–40.

Further Reading

There are no textbooks which describe folds and folding in detail including the recent development in the field. Hobbs, B. E., Means, W. A. and Williams, P. F., 1976, *An Outline of Structural Geology*, 571 pp., John Wiley, New York, is an excellent introduction to structural geology. It contains a chapter on folds, but is insufficient for an in-depth study of folding. In order to learn the structural features of various types of folds, individual papers such as those quoted in the text have to be referred to. Biot's (1965a*) book is recommended for studying mechanical theories of folding. The mechanical basis to read his book will be acquired in: Jaeger, J. G. 1968, *Elasticity, Fracture and Flow*, 268 pp. Methuen, London.

Constitutive equations (stress–strain relationships) of rocks are treated concisely in Chapters 1 and 2 of Volume 2 of the Iwanami Earth Science Series. Johnson (1977*) summarizes a series of recent theoretical analyses of folding by himself and his co-workers, but their results are not compared with the results from previous theories and with the characteristics of natural folds in sufficient detail. Ramberg (1967*) gives an excellent treatise, both theoretical and experimental, on the effects of gravity.

Geological Structures
Edited by T. Uemura and S. Mizutani
©1984 John Wiley & Sons Ltd.

6

The flow of Rocks

Phenomena in which solid materials flow are not really unusual and many examples can be seen in nature, for instance the flow of glaciers. It is well known that even if something as hard as rock is subjected to sufficiently large forces for sufficiently long it will finally begin to flow. It is also obvious that the development of many geological structures has been related directly or indirectly to the flow of rocks, the best example of this being folding. The tremendous variety of folds seen throughout orogenic belts demonstrates that the rocks and strata have bent like jelly in response to huge forces and have not returned to their original state. Folding mechanisms were the subject of detailed discussion in the preceding chapter and such 'permanent deformation', i.e. deformation which does not return to its original state even though the external force has ceased to operate, is in fact one of the particular features of the phenomena known as 'flowing'.

In broad terms we may identify two principal types of phenomena with differing origins both of which constitute rock flow. The first is deformation of unconsolidated sedimentary material or the growth of *salt domes*, etc., whereas the second derives from developmental processes for geological structures such as folds, etc. The difference between the two lies in the fact that whereas the first type is generated in a field where basically only gravity is at work, the second type appears in fields where so-called *tectonic stresses* exist in addition to gravity. The second type of process is known as '*tectonic flow*'. The problem of tectonic flow is one of the most important topics in structural geology. The phenomena themselves are complex and many substantial aspects thereof are still not fully understood. In this chapter we describe the basic features of tectonic flow, citing several examples of its application to the mechanical analysis of geological structures.

6.1 Tectonic flow and its characteristics

The study of the deformation and flow of materials is known as *rheology*. Deformation phenomena may first be broadly classified into *elastic deforma-*

tion and flow, and the latter is further divided into *plastic* and *viscous flow* depending on the presence or otherwise of a yield value. Phenomena of flow is a continuous deformation in which strain advances over time under a stress state and a *permanent strain* will remain even if the stress is subsequently removed. In contrast to this, strain in elastic deformation is independent of time and can appear almost instantly in response to a stress. When the stress is removed the strain recovers completely in an equally rapid manner.

Mechanical models (rheology models) are convenient to any description of deformation characteristics. The idea of the mechanical model is to express the properties of any material by combining three models, namely the *spring* for a perfectly elastic material rules by Hooke's Law (*Hookean elastic body*), the dashpot for an ideal viscous material ruled by Newton's laws of viscosity (*Newtonian fluid*) and the *latch* or *slider* to express the presence of a yield point (Figure 6.1)

It hardly needs to be said that the materials which constitute geological structures are aggregates of minerals we term 'rocks'. Generally the flow characteristics of rocks clearly exhibit the behaviour of plastic bodies with yield points, but this behaviour is known to be often highly complex ranging from close to that of an elastic body to close to that of a viscous body depending on the deformation conditions. It is therefore far from easy to assemble a mechanical model able to explain fully the deformation properties of a rock in all situations.

Be that as it may, it is still convenient when describing phenomena of flow to use rheology equations to express the relationships between stress, strain

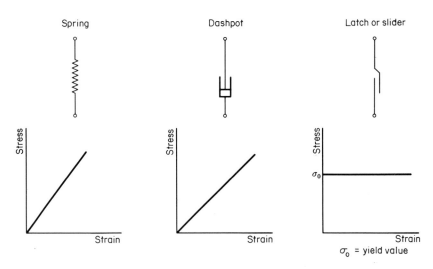

Figure 6.1   The three elements of mechanical models and their characteristics

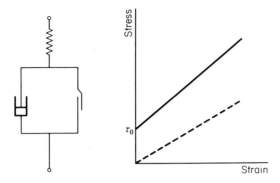

Figure 6.2 Bingham model and its stress–strain rate curves. (Under constant stress in excess of yield point τ_0). Broken line $\ominus$ Newtonian fluid

and time using rheology constants as parameters such as viscosity coefficients. For example, in laminar flow of Newtonian fluids the relationship between the strain rate $\dot{\epsilon}$ and the stress τ taking viscosity η as a constant is given by:

$$\dot{\epsilon} = \tau/\eta$$

The most straightforward plastic body widely used as a model for rock flow is the *Bingham plastic body* illustrated in Figure 6.2. A rheology equation showing the behaviour of this system under constant stress is called the *Bingham* equation and is written as follows:

$$\dot{\epsilon} = (\tau - \tau_0)/\eta'$$

η' is plastic or pseudo-viscosity, τ_0 is the yield value and both are constants. These two types of flow maintain a linear relationship between the stress and the *strain rate* as one can see from the form of the equations and essentially they possess the same properties. An important point is that phenomena of 'flowing' are inseparable from 'time'. This fact is expressed by the inclusion of time in the above rheology equations in forms such as the strain rate ($\dot{\epsilon} = \partial\epsilon/\partial t$) or viscosity coefficient (the dimension of η is $ML^{-1}\,T^{-1}$) etc.

Phenomena of flow are determined by three factors, i.e. stress, strain rate and viscosity (or an equivalent measure) but only two of these are independent of each other in linear flow. Let us examine this for the case of tectonic flow. Imagining the stress field first, only the volume changes and no deformation will occur in a hydrostatic stress field only. When an arbitrary stress field is expressed as three principal stresses the difference between each principal stress and the average of the three principal stresses is called the '*deviatoric stress*'. On the other hand, the mean value of the three principal stresses shows the hydrostatic part of this stress system. Because deformation phenomena including flow in general only become possible in the presence of deviatoric stress, one may say that the essence of the so-called 'tectonic stress field' lies in the existence of deviatoric stress, i.e. in the triaxial stress state. Although being able to make qualitative statements, for the purpose of any quantitative treatment, we have no means of directly estimating the values for the deviatoric stresses which have produced rock flow throughout geological time.

On the question of strain rate, an average value may be calculated if the time factor is known, because we can sometimes obtain the strain itself from surveys and measurements of the deformed structures. However, time obtained by methods of stratigraphy of historical geology is not in general sufficiently accurate to be useful in this kind of discussion. If measurements of only viscosity coefficients were possible in such cases some quantitative discussion of the absolute values of tectonic stresses might become feasible with respect to phenomena throughout geological time.

To apply such ideas on tectonic flow to an elucidation of the formative mechanism of geological structures one must not only carry out sufficient surveys of the structures but also reach some qualitative conclusions as to their origins.

6.2   *Creep* and stress relaxation

We now mention creep and stress relaxation which of all the mechanical properties of rocks concerned with flow are thought to be of particular importance in the formation of geological structures.

'Creep' is defined as the phenomenon in which the strain in an object increases over time when it is subjected for a long period to a constant stress in excess of a certain magnitude (yield value). The '*creep curve*' expresses this as a relationship of strain to time. Creep may generally be divided into three stages. The first stage is known as primary or transitional creep. It follows instantaneous elastic deformation, and the strain rate decreases over time to produce an upward convex creep curve. The second stage is secondary or steady creep in which the strain rate is constant and the creep curve is a straight line. During the third stage cracks appear and develop rapidly so that the strain rate increases and the object finally ruptures. This is called tertiary or accelerating creep (Figure 6.3). Much theoretical and experimental work has been done on creep in single mineral crystals or mono-mineralic rocks such as limestone.

The phenomenon of creep can be recognized to a certain extent in rocks and *Burgers model* is commonly used to explain primary and secondary creep

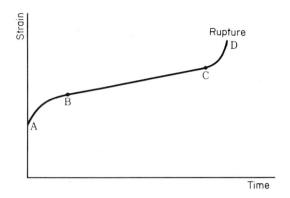

Figure 6.3   Creep curve. AB — primary creep; BC — secondary creep; CD — tertiary creep

(Figure 6.4). This is combination of the Maxwell and Voigt (*or Kelvin*) *models* in series and when a constant stress σ_0 is applied to this system the relationship between strain ϵ and time t may be given as follows:

$$\epsilon = \sigma_0 \left[\frac{1}{\gamma_M} + \frac{t}{\eta_M} + \frac{1}{\gamma_V} \left\{ 1 - \exp\left(-\frac{\gamma_V}{\eta_V} t \right) \right\} \right]$$

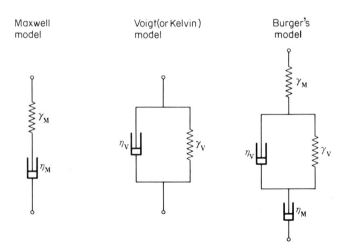

Figure 6.4   Burgers model and elements thereof

γ and η show the elastic modulus and the viscosity coefficients and the subscripts M and V refer to the *Maxwell* and *Voigt models*. At $t = 0$ this equation expresses elastic deformation as $\epsilon = \sigma_0/\gamma_M$, and at $t = \infty$ it shows secondary creep with a constant strain rate given by $\dot{\epsilon} = \sigma_0/\eta_M$. Whilst t is small it can express primary creep in which $\dot{\epsilon}$ is gradually decreasing. In addition to the above, various forms of equations such as exponential functions, logarithmic function and hyperbolic functions have been proposed to express creep strain and stress.

Similarities to rock creep as a phenomenon may be found to a certain extent in the creep of minerals, but the actual mechanisms of *rock creep* are far more complex than those for minerals. Notwithstanding, rock and strata creep is thought to play a principal role in the formation of geological structures for which the passage of vast lengths of time is required.

Stress relaxation is another important factor in this. It is generally seen in the Maxwell model or compound models containing it, and is described as the phenomenon in which stress gradually decreases when strain is held constant. Taking Young's modulus of the spring as E, the viscosity of the dashpot as η, stress as σ and strain as ϵ, then the rheology equation for the Maxwell model shown in Figure 6.5 may be written as follows:

$$\dot{\epsilon} = \frac{\dot{\sigma}}{E} + \frac{\sigma}{\eta}$$

If strain in the system is now held constant and $\sigma = \sigma_0$ when $t = 0$, stress is expressed by

$$\sigma = \sigma_0 \exp\left(-\frac{E}{\eta}t\right)$$

and decreases exponentially with time. Taking t_e as the time needed for the stress to reach $1/e$ of the initial stress (e is the natural logarithmic base) then

$$t_e = \eta/E$$

This is called the *relaxation time*.

Let us consider the tectonic implications of stress relaxation. Griggs (1939*) performed high pressure triaxial tests on Solenhofen limestone and calculated 2.2×10^{22} poise as the value corresponding to its viscosity coefficient using data from creep tests held for 550 days at 23 °C and with a differential stress of 1400 kg/cm^2. He suggested this would probably indicate a minimum level for the average value for the rocks which constitute the upper crust. This value corresponded well with the the viscosity coefficient obtained by Gutenberg (1941) in a simulation of the postglacial Fennoscandian

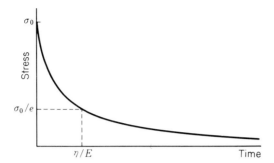

Figure 6.5 Stress relaxation curve and relaxation time in the Maxwell model

upheavals using the Maxwell model. Now, if $\eta = 2.2 \times 10^{22}$ poise and $E = 7 \times 10^5$ kg/cm^2 then $t_e \doteqdot 1 \times 10^6$ (years). Price (1959) noticed that this value for η was (a) for a rock which flows easily, i.e. limestone and (b) that it would be affected by temperature at depth. He thus deduced that the stress relaxation time of rocks in the upper crust would not be less than 10^6 years and would perhaps exceed 10^7 years in which case the time required for the stress to fall to 1 per cent of the original would be nearly 5×10^7 years.

If this inference is correct it means that a stress acting on a certain part of the crust at a certain time will subsequently continue to exist for an extremely long period as a so-called *residual stress*. For example, the three major cycles of orogenic movements throughout the earth since the Palaeozoic era (namely Caledonian, Variscan and Alpine orogeny) have all been spaced at intervals of about 1.5×10^7 years and subsequent ones will originate in a situation in which *tectonic stresses* from preceding orogenic periods are probably still very much in evidence. This is worth noting for the mechanical aspects of tectonic history, but there are virtually no examples of concrete studies of this. However it will probably be an important field of research in future.

6.3 Tectonic stress fields and the growth of folds

The flow of rocks to form geological structures is generally extremely slow. If one were to express the strain rate as how much distortion occurred in one second (e.g. if it were 50 per cent it would be 0.5/s = 5×10^{-1}/s, or if 10 per cent in one million years, 3.3×10^{-15}/s) then the value would perhaps be in the order of 10^{-15} to 10^{-16}/s. It is difficult to imagine flows at such ultra-low strain rates as always being continuous but nevertheless one may suppose creep or a flow with similar characteristics. Experiments conducted over long periods of time have also shown that the yield point even of granites is extremely low, and that most may be regarded as viscous fluids (Kumagai and Ito, 1968*). Combining these depths means that one can make hypotheses about secondary creep at various depths as a model of tectonic flow in rocks. As mentioned earlier secondary creep is a steady flow at a uniform strain rate. Furthermore, the most important point on which flow differs from elastic deformation is its time dependent nature. This ultimately comes back to the problem of establishing viscosity coefficients.

Growth of geological structures due to rock flow means the progress of plastic flows in a tectonic stress field. Therefore the key to solving this problem lies in measuring the viscosity coefficients of rocks under high confining pressures. Almost the only way to do this is by creep tests. It may be useful to use a so-called triaxial testing apparatus of the two-axial equal pressure type ($\sigma_1 \neq \sigma_2 = \sigma_3$) in which the confining pressure in a high pressure vessel is controlled by variations in the volume of a fluid, and a differential stress $\sigma_1 - \sigma_2$ is generated by axial pressure loading from one direction. This is also a model experiment for tectonic flow. Actual rock is used as the sample and the stress field under which the tectonic flow occurs is defined as the experimental conditions of confining pressure and differential stress (including temperature if necessary). The experiment is continued until a state of steady creep is verified and the viscosity coefficient is then obtained from the slope of the creep curve. The viscosity coefficient obtained in this way is that for normal strain (λ) and is not the viscosity coefficient η for shear strain such as laminar flow. If the sample is incompressible the relationship $\lambda = 3\eta$ exists and so if differential stress is taken as $\sigma_1 - \sigma_2$ then:

$$\eta = (\sigma_1 - \sigma_2)/3\dot{\epsilon}$$

Experiments like these were initially performed by Griggs (1936*, 1939*) on *Solenhofen limestone*, etc. For example, he calculated 1.29×10^{15} poise as the equivalent viscosity with a differential stress of 5460 kg/cm^2 and confining pressure of 10 000 kg/cm^2 (Figure 6.6). Creep tests along similar lines using triaxial testing apparatus have also been performed on mudstones of the Tertiary *Nishiyama formation* which made the folds in the oilfields of Niigata. Several values for equivalent viscosities have been calculated and so let us examine their relationship to the growth of folded structures (Uemura, 1976*).

In these tests the geological situation inferred was of folds at a depth of about 1500 m, and so a confining pressure of 300 kg/cm^2 was selected. Some of the results are shown in Table 6.1 and surprisingly it was found that steady creep advanced with a strain rate in the order of 10^{-8}/s at tectonic stress fields represented by differential stresses of 100 – 500 kg/cm^2. The equivalent viscosities calculated from this were in the order of 10^{15} poise. In the case of

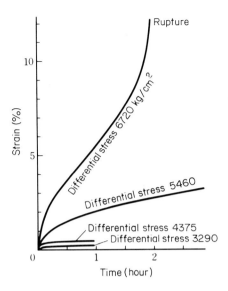

Figure 6.6 Creep curve for Solenhofen limestone under $10\,000$ kg/cm² confining pressure (Griggs, 1936[*])

Table 6.1 Creep test results for Nishiyama mudstone and values calculated from them. Tests at confining pressure 300 kg/cm² and room temperature. Bottom two lines are extrapolated values from the relationship of differential stress and strain rate (Uemura, 1976[*])

Differential (tectonic) stress (kg/cm²)	Strain rate (s)	Viscosity correspondence value (poise)	Time required to reach 30% strain (poise)
500	5.9×10^{-8}	2.8×10^{15}	0.16
400	4.7×10^{-8}	2.8×10^{15}	0.20
300	2.5×10^{-8}	3.9×10^{15}	0.38
200	2.2×10^{-8}	3.0×10^{15}	0.43
100	8.9×10^{-9}	3.7×10^{15}	1.07
20	1.0×10^{-10}		95.1
2	1.0×10^{-13}		9.5×10^{4}

water-saturated test pieces, values several times larger were obtained. We can use these values to discuss problems of strain rates, growth of folds and tectonic stress fields. Looking at Table 6.1, strain reached as much as 30 per cent in just over a year when the *tectonic stress* was 100 kg/cm^2, but such enormous changes are obviously not realistic. There are two principal lines of thought about the development of folds in the oilfield of Niigata. The first is the view represented by Omura (1930*) which emphasizes the existence of a 'folding phase' whereas the other, represented by Kanahara, does not accept such a phase but claims that the folds have developed gradually from the time of the initial layer deposition through the present day. According to the more recent ideas of Suzuki (1971*) there was a period of rapid fold growth in the middle Pleistocene but it seems that gentle folding movements beginning earlier still in the late Miocene period also took place. If we accept this idea and assume that most of the folds grew in the middle Pleistocene, this gives us a folding period of 100 000 years at most. As it seems reasonable to suppose that the mechanical properties of the Nishiyama formation, which belongs to the lower Pliocene series, were not so different than they are today, the results of tests performed on current samples are directly relevant.

Another important factor in expanding this discussion is to estimate the value of the tectonic stress. With respect to the stresses operative in the present crust, Chinnery (1964*) reported that the stress drop associated with major strike-slip faulting and earthquakes was about 10–100 bar. Hast (1967*) also stated that measurements of absolute crustal stresses in Scandinavia indicated a total horizontal stress at the surface of 180 kg/cm^2 and that this value increased in proportion to the depth reaching 1000 kg/cm^2 at a depth of 1000 m. Herget (1974*) reduced all the data for Canada and reported the average horizontal stress to be (83 ± 5) kg/cm^2 + (0.407 ± 0.023) H kg/cm^2 (H = depth in metres). Fujii and Ito (1973*) believed that a value of about 180 bar was reasonable for tectonic stress acting on the then crust from the formative mechanism of sedimentary basins in the upper Pliocene Otadai formations distributed in the Boso Peninsula. Ito *et al.* (1976*) demonstrated that maximum compressive principal stress was almost horizontal in an E–W direction from measurements of bedrock stress in an underground power station in Omachi, Nagano Prefecture, and found its value to be 100–180 kg/cm^2. All this research would seem to indicate that values of tectonic stress in the upper crust are in the order of 10–100 kg/cm^2.

However, as mentioned earlier, the results for the Nishiyama formations were obtained in case of a differential stress of 100 kg/cm^2, but even assuming a differential stress of 20 kg/cm^2 the strain still reached 30 per cent in 95 years as shown in Table 6.1. If strain is assumed to have reached 30 per cent over 100 000 years then the tectonic stress continuously at work during that time would have a differential stress value of no more than 2 kg/cm^2. This value seems to be far too small, but could perhaps provide one solution. Another solution is the idea that the stress field in which the Nishiyama folds grew still had a differential stress level of about 100 kg/cm^2 but that it acted intermittently rather than constantly. In this case to reach the strain of 30 per cent in 100 000 years a differential stress of about 100 kg/cm^2 need act for only about one year in total. This interpretation is quite close to the idea that folds develop intermittently when major earthquakes occur. Assuming that one major earthquake happens every 1000 years, then one hundred would occur in 100 000 years and a differential stress of about 100 kg/cm^2 would need to exist for only a few days, perhaps preceding each occurrence.

Whereas the first approach postulates that folds grow slowly as a result of creep, which progresses continuously due to an ever present minute differential stress, it is of great interest that the second interpretation, which assumes interrupted growth periods of only extremely short duration just before major earthquakes even given the same creep, is closely related to seismicity in geologic time. Furthermore, to create a folded structure in a specific direction, the principal stress axis must be in a continuously uniform direction even if the differential stress is constantly very weak, at only a few kg/cm^2, or whether it reaches a value of about 100 kg/cm^2 on only a few days every 1000 years.

The above discussion has been limited to the problems of stress in the Nishiyama formation but in fact we are still left with the problem of what were the significant factors which resulted in such a stress field? To take the discussion any further we would have to examine the relationship between the whole Tertiary system containing the Nishiyama formation and the basement rocks. As, if we could construct a model in which, despite a differential stress of 10–100 kg/cm^2 in the basement, a differential stress of no more than a few kg/cm^2 was generated in the overlying strata, this would give a concrete basis to the first solution.

6.4 Mechanisms of boudin formation

It is rare for rocks and strata in nature to be homogeneous. Their normal state in fact is to consist of inhomogeneous complex bodies. Therefore portions which will and will not flow under certain conditions coexist in a series of rocks, and the former areas may apply pressure to the latter and fracture them. Structures formed in this way include those called '*boudins*' or '*boudinage*'. This is a French word meaning a sort of sausage.

Generally speaking boudins are formed when rocks, which flow with relative difficulty, form blocks inside rocks which flow more readily. The cross-section of such blocks is nearly oblong and form relatively close to their original positions. The ends of the oblong may be elongated into lens shapes due to plastic deformation, or they may take the form of so-called *pinch and swell* structures as if connected to an adjacent lens. Both types are thought to

Figure 6.7   Progressive development of boudinage structure in calc-silicate bands in marble (Ramsay, 1967)

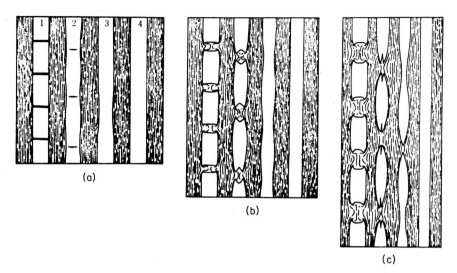

Figure 6.8   Development of boudins and pinch and swell structures. Boudin development progresses in the order (a), (b), (c). 1, 2, 3 and 4 are all competent layers, and the smaller the identification number the greater the competency of layer (Ramsey, 1967).

be attributable to the same cause. Several examples of boudins are shown in Figures 6.7 and 6.8.

Boudins have been studied by many geologists since Lohest (1909*) first described them in Belgian sandstone and gave them their name. More recently experimental and theoretical work has been carried out as well as descriptions and classifications of them. In particular, mathematical analysis of their mechanism have been established by Ramberg (1955*) and Gzovsky (1960*). Like folds, they are very useful and important structures in determining the mechanical conditions prevailing at the time of formation. Appropriate mechanisms of formation have been imagined for each of the several types of boudin. In this section we deal with the most basic types.

When a system of alternating incompetent and competent strata is compressed at right angles to the stratification, the competent layers are pulled by the flow of incompetent layers. If the competent layers are capable of some plastic deformation they will develop pinch and swell or lens shaped structures dependent on their degree of competency. If brittle tensile fractures occur then typical boudins will be formed (Figure 6.9). Ramberg (1955*) assumed that viscous flow would characterize the flow of the incompetent layers and the behaviour of the competent layer up to fracturing

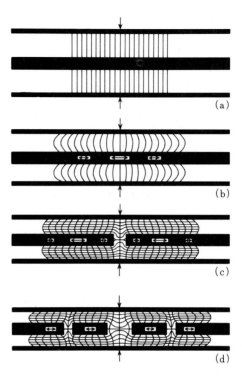

Figure 6.9   Schematic diagram showing processes of boudin formation (Ramberg, 1955*)

would correspond to elastic deformation, and he proposed the following theoretical conditions for the formation of boudins.

$$\frac{z_b}{z_d}S_t > P_0 - P_l > \frac{1}{2}\frac{z_b}{z_d}S_t$$

z_b and z_d are the thicknesses of the competent and the incompetent layers, P_0 and P_l are the pressures at the centre and at the ends of the competent layer of length $2l$, S_t is the tensile strength of the competent layer. In brief when the value of $P_0 - P_l$ is within the range shown by this inequality, a boudin of length $2l$ will be formed. $P_0 - P_l$ is given by the following equation:

$$\frac{\partial z_d}{\partial t} = \frac{z_d{}^3}{6l^2\eta}(P_0 - P_l)$$

η is the viscosity of the incompetent layer, and $\partial z_d/\partial t$ is the rate of reduction of the layer thickness. Ramberg calculated that $\partial z_d/\partial t \doteq 10^{-11}$ cm/s assuming that $P_0 - P_l = 50$ kg/cm^2, $z_d = 100$ cm, $l = 75$ cm and $\eta = 10^{20}$ poise. As this is equivalent to about 3 cm in 10 000 years he felt it indicated a reasonable rate at which layers could be compressed in reality, but there is no particular proof for his estimates of $P_0 - P_l$. Let us now consider an example in which Ramberg's analysis is expanded and applied to actual boudins observed in the field (Uemura, 1965*, 1971b*).

The southern part of the Shimanto terrain in the Kii Peninsula is called the Muro terrain, where the Palaeogene to the early Miocene Muro group is distributed. The predominant rocks which constitute the Muro group are an alternation of sandstone and shale, and the sandstone sometimes forms boudins. In brief, the sandstone and the shale layers correspond respectively to the competent and the incompetent layers. Geological surveys were undertaken over a wide area, and the dimensions of a number of boudins were measured. Average values were 28 cm, 12 cm and 15 cm for length ($2l$) and thickness (t_b) of boudin, and thickness of the shale (t_d) between the boudins. Taking the average value of $\partial z_d / \partial t$, having eliminated $P_0 - P_l$ which cannot be measured, one obtains from the above two equations of Ramberg

$$\frac{\partial z_d}{\partial t} \fallingdotseq \frac{S_t}{\eta} \cdot \frac{z_d^2 z_b}{8l^2}$$

Because field observations revealed that the amount of deformation had not been particularly large, if one regards $z_b = t_b$ and $z_d \fallingdotseq t_d$, the value of $\partial z_d / \partial t$, relating to the flow rate of the shale which formed the boudins, is determined by the two physical constants S_t and η.

S_t is the tensile strength of the sandstone at the time when the boudins were formed and so measured values from present samples cannot be used as they are. Reduction of data of the measured values of Japanese sandstone from each geological age has permitted the construction of a graph showing the approximate relationship between its strength and age. From this graph an estimated value of 10–40 kg/cm^2 was obtained for the time of boudin formation (it might be the same as the folding) in the Muro group. A comparable value of 30–50 kg/cm^2 was obtained using the results of radial compression tests of irregularly shaped pieces of the Muro sandstone. The final element remaining is the viscosity coefficient η. Assuming $\eta = 10^{22}$ poise as various studies on the flow of crustal materials have suggested $\partial z_d / \partial t$ as the strain rate gives a value of about 1–0.3×10^{-14}. If shale 10 cm thick flowed at a rate which made it become $3 - 1$ cm thinner in 1 million years boudins resembling those actually observed should be formed.

On the other hand, because the folded Muro group is covered obliquely and unconformably by the middle Miocene system, the period of folding is limited to a few million years between the early and middle Miocene. The *strain rate* obtained above is not inconsistent with this fact. However, the value of 10^{22} poise may be thought perhaps to show an upper limit for the viscosity of sedimentary rocks and so, if $\eta = 10^{21}$ poise, a strain rate larger by a factor of ten would be obtained, and the folds of the Muro group would have been completed within a few tens of thousands of years. Conversely, if we were actually able to determine the duration of the period of folding, we could calculate the viscosity. In general, however, we have no such method. Any quantitative discussion of this problem depends on being able to obtain an accurate value for the viscosity coefficient. If this could be resolved then the mechanical conditions prevailing at the time of boudin formation, such as the magnitude of the forces which brought about the flows of shale and the pressure gradient in the direction of flow, could be clarified. Any mechanical analysis of geological structures formed by the tectonic flow of rocks and strata finally comes down to the relationships between strain rate, viscosity coefficients and stress. Problems concerned with the mechanisms of the development of boudinage structures are no exception to this.

Takeshi Uemura

References

Chinnery, M. A. (1964): The strength of the earth's crust under horizontal shear stress, *J. Geophys. Res.*, **69**, 2085–9.

Fujii, K. and Ito, H. (1973): Tectonic stress and viscosity of the earth's crust at the time of deposition of the Otadai formation in the Boso Peninsula, *Japan. J. Geol. Soc. Japan*, **76**, 89–98.

Griggs, D. T. (1936): Deformation of rocks under high confining pressures, *J. Geol.*, **44**, 541–77.

Griggs, D. T. (1939): Creep of rocks, *J. Geol.*, **47**, 225–51.

Gzovsky, M. V. (1960): Tektonofizika i problemy strukturnoi geologii, *Mejdunarod. geol. kongr., 21 Sess., Doklady Soviet geolog., Prob.* 18, 17–31.

Hast, N. (1967): The state of stresses in the upper part of the earth's crust, *Eng. Geol.*, **2**, 5–17.

Herget, G. (1974): Ground stress determinations in Canada, *Rock Mech.*, **6**, 53–64.

Ito, H. *et al.* (1976): The contracting Japanese Islands — rock mechanics seen from experiments, measurements and in the field. *Kagaku*, **46**, 745–54.

Kanehara, K. (1950): Geology of the oilfield of Niigata (II). *J. Japan. Assoc. Petroleum Technologists*, **15**, 62–83.

Kumagai, N. and Ito, H. (1968): Results of experiments of secular bending of big granite beams extending for 10 years and their analyses. *J. Soc. Materials Sci. Japan*, **17**, 925–32.

Lohest, T. (1909): De l'origine des veines et des geodes des terrains primaires de Belgique, *Ann. Soc. Geol. Belgique*, **36**, B275–282.

Omura, K. (1930): Geology and petroleum deposits of the oilfield of Echigo, *J. Geol. Soc. Japan*, **37**, 775–91.

Price, N. J. (1959): Mechanics of jointing in rocks, *Geol. Mag.*, **96**, 149–67.

Ramberg, H. (1955): Natural and experimental boudinage and pinch-and-swell structures, *J. Geol.*, **63**, 512–26.

Suzuki, *et al.* (1971): On the mechanisms of folding in the Niigata Tertiary basin. *J. Geol. Soc. Japan*, **77**, 301–15.

Uemura, T. (1965): Tectonic analysis of the boudin structure in the Muro group, Kii peninsula, southwest Japan, *J. Earth Sci., Nagoya Univ.*, **13**, 99–114.

Uemura, T. (1971a): Some problems on the tectonic flow of rocks. *J. Geol. Soc. Japan*, **77**, 273–8.

Uemura, T. (1971b): On the two types of tectonic lenses and their co-existence. *Earth Sci. (Chikyu Kagaku)*, **25**, 30–35.

Uemura, T. (1976): Some problems on earthquake and the formation of geologic structures. *Mem. Geol. Soc. Japan*, **12**, 43–9.

Further Reading

The following are specialized works dealing theoretically with the flow of solids and rocks:

Jaeger, J. C. (1969): *Elasticity, Fracture and Flow*, 3rd ed., 268 pp. Methuen, London.

Jaeger, J. C. and Cook, N. G. W. (1969): *Fundamentals of Rock Mechanics*, 515 pp., Methuen, London.

Nadai, A. (1950): *Theory of Flow and Fracture of Solids*, 2nd ed., 572 pp., McGraw-Hill, New York.

Ueda, S. (ed.) (1974): *Flow in Solids*, 324 pp. Tokai Univ. Press, Tokyo.

The following works consider boudin formation and its associated problems:

Gay, N. C. (1968): The motion of rigid particles embedded in a viscous fluid during pure shear deformation of the fluid, *Tectonophys.*, **5**, 81–8.

Gay, N. C. (1968): Pure shear and simple shear deformation of inhomogeneous viscous fluids. 1. Theory, *Tectonophys.*, **5**, 211–34.

Smith, R. B. (1975): Unified theory of the onset of folding, boudinage, and mullion structure, *Geol. Soc. Amer. Bull.*, **86**, 1601–9.

Smith, R. B. (1977): Formation of folds, boudinage, and mullion in non-Newtonian materials, *Geol. Soc. Amer. Bull.*, **88**, 312–26.

Stromgard, K. E. (1973): Stress distribution during formation of boudinage and pressure shadow, *Tectonophys.*, **16**, 215–48.

Umemura, H. (1972): Origin of boudin structure in Gosaisho-Takamuki metamorphic rocks, Abukuma Plateau. *Res. Rept. Kochi Univ.*, 14, *Nat. Sci.*, 209–27.

The following works deal with rock creep and its associated problems:

Heard, H. C. (1963): Effect of large changes in strain rate in the experimental deformation of Yule marble, *J. Geol.*, **71**, 162–95.

Heard, H. C. and Raleigh, C. B. (1972): Steady-state flow in marble at 500° to 800° C, *Geol. Soc. Amer. Bull.*, **83**, 935–56.

Rutter, E. H. (1972): On the creep testing of rocks at constant stress and constant force, *Int. J. Rock Mech. Min. Sci.*, **9**, 191–5.

Rutter, E. H. and Schmid, S. M. (1975): Experimental study of unconfined flow of Solenhofen limestone at 500° to 600°C, *Geol. Soc. Amer. Bull*, **86**, 145–52.

Scheidegger, A. E. (1970): On the rheology of rock creep, *Rock Mech.*, **2**, 138–45.

Geological Structures
Edited by T. Uemura and S. Mizutani
©1984 John Wiley & Sons Ltd.

7

Fractures and Bedrock

The structures seen in rocks are a means of assessing what sort of mechanical history led to their creation and the rock units themselves. Most research on geological structures is undertaken from such a viewpoint, and this book too has been dealing with this kind of study. Structures, such as fractures, found in rock are of major concern not merely in terms of the interest physical science has in their derivation, but also because Man makes use of them. The mechanical properties, which the structure or formation possesses at present are, therefore, far more important than understanding the origin of the structure, and information about its derivation can, and should, be usefully applied to understand it. If one also considers 'current properties' from the point of view of the use of structures, the results may also contribute to our understanding of their origins.

In this chapter we have a complete change of emphasis, and discuss some problems of structures in terms of engineering geology or *geotechnics* in an attempt to give an interpretation with which no existing textbooks on structural geology are concerned. One cannot deny that a great gulf currently exists between structural geology and engineering geology. However, readers of this chapter should at least become more aware of the fact that, notwithstanding this gulf, the interrelation between the two is essentially vital for both, and that this opens up a completely new field of study.

It is with such expectations that we discuss the techniques and concepts of engineering geology, citing several examples, in what is very much a developing subject. Lack of space precludes a sufficiently thorough discussion, but in this chapter we grapple particularly with the problems of how to assess the mechanical properties of rock masses containing joints and faults and how one should deal with them in practice. The future of engineering geology, which is directly connected with structural geology, will have to be created through the efforts of both disciplines. We hope that people with a particular interest in engineering will refer to the additional reading at the end of the chapter. We hope also that this chapter will be seen as an extension of the objectives of the book as a whole, rather than its inadequacies as a discussion of structural geology.

7.1 Fractures and engineering geology

Let us introduce an example of the treatment of faults and fractures, which are frequently encountered in textbooks of engineering geology and rock mechanics, to aid our understanding of the engineer's view of rock masses.

Figure 7.1 is a horizontal profile showing the distribution of faults and fractures around the site of a certain arch dam. The dam is designed so that the load from the embankment is supported mainly by the lateral bedrock masses. One must, therefore, prepare various horizontal profiles every few tens of metres to show the geological and physical properties of the rock. Let us presume that a fault and fracture system such as shown in Figure 7.1(a) is

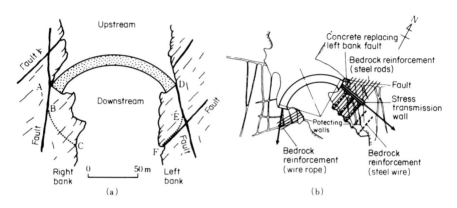

Figure 7.1   Arch dam and cracks in bedrock (Doboku Gakkai, 1975*). (a) Effect of faults on bedrock stability (see Table 7.1). (b) Plane diagram of bedrock works around Kawamata Dam (910 m level)

found to predominate. A two-directional fault system lies on the left bank of the planned abutment for the dam and these directions correspond with the direction of slip planes which would easily be generated in the rock as a result of loading from the dam wall. Generally speaking *shear strength* along a fault plane is markedly less than that of the surrounding rock mass. Therefore, the rock mass might fracture along this pair of faults. To verify this, the shear strengths of the faulted areas and the surrounding rocks are tested (Table 7.1). If *stability calculations* for *shear fracturing* are made using these values, then, as expected, there will be a slip plane along the fault, on the left bank, with the design shown in Figure 7.1(a). Thus the ratio between the *shearing resistance* of the rock mass and the operational force from the dam wall, i.e. the safety factor, will clearly be lower than on the right bank. A value of four or more is deemed to be essential, but on these results it would be impossible

Table 7.1   Mechanical properties of the bedrock and its stability against load from the arch dam wall (see Figure 7.1)

	Right bank		Left bank	
Stress applied from arch dam wall to side walls	1330 t/m^2		1330 t/m^2	
Shear resistance of each rockmass in Figure	5 t/m^2	(A–B)	5 t/m^2	(D–E)
7.1(a) and along faults	100 t/m^2	(B–C)	5 t/m^2	(E–F)
Friction coefficient of each rockmass in Figure	0.6	(A–B)	0.36	(D–E)
7.1(a) and along faults	1.0	(B–C)	0.36	(E–F)
Safety level relative to shear failure	6.0		1.1	

to build the dam unless the above safety factor could be raised in some way. Other possible locations for the dam site are then investigated and a comparative study of each location is made, on the basis of which the safest and most economical site is finally chosen. This type of site evaluation can be illustrated by the Kawamata Dam shown in Figure 7.1(b). On this site it was decided that the best course of action would be to improve the *bedrock* on the left bank by building a concrete wall, such as that shown in the diagram and to disperse stress from the dam wall into the *bedrock*. The required degree of safety was achieved by 'sewing up' the cracks with wire rope or steel rods. In this example, the work of the geology specialist involves:

(a) Preparing maps showing the precise geological composition of the limited area of the dam site and an accurate distribution of faults and fractures in the bedrock,
(b) Assessing the strength of the faults or rock masses containing fractures,
(c) Co-operating with engineers in predicting how the bedrock will behave when subjected to the load from the dam, and to devise counter-measures including alteration of the site.

   In this chapter we discuss, from such a point of view, how one can understand the properties and distribution of non-uniform rock masses containing faults and fractures, and how one can predict their behaviour when subjected to natural or man-made forces, purely on the basis of their mechanical properties.

7.2   *Discrete surfaces* in bedrock

When considering 'rock' in engineering geology this means a block to be subjected to engineering works, and the terms 'bedrock' and 'rock mass' are in common use. Bedrock, quite obviously, is neither a uniform nor a continuous material. It usually contains planes of discontinuity such as faults or joints. Because such discontinuous planes have a considerable effect on the mechanical properties and the permeability of the bedrock, they should be taken into account in any study of the rock or the bedrock.

Many things may be classed as discontinuous planes in terms either of their derivation or configuration. There are no established rules for their classification and this will vary subjectively from person to person. In this section we call all types of planes of discontinuity, e.g. cracks due to bedding, schistosity, joints, fissures, fault shatter zones and seams, etc. by the one term, namely 'discrete surfaces in a geologic body'. When there is no need for particular classification we use the term 'fractures' to represent planes of discontinuity. *Joints* are regular and sometimes regional *fractures*, fissures are irregular local fractures, whereas *seams* are thin soft weak layers in the hard bedrock such as narrow shatter zones or weathered zones or deformed veins.

Many discrete surfaces in the bedrock are thought to have developed in association with the origins of the rocks or their condition when produced. In plutonic rocks such as granite, lattice-shaped groups of joints or sheeting joints following topographical planes are particularly marked, whereas one finds columnar or sheeting joints in lava, welded tuff and dykes, bedding planes and discontinuous groups of joints, intersecting them, in sedimentary rocks and schistosity and groups of joints at right angles in crystalline schists. Generally speaking, fractures formed as a result of extension have rough planes and open out near the surface, whereas cracks, formed as a result of shearing, are frequently accompanied by shatter zones and are smooth and in tight contact. Minerals which slip readily such as graphite, talc or chlorite are often arranged in discontinuous planes of schistosity or bedding, and the mechanical properties of the rock may be considerably affected by them.

7.3   Changes in the physical properties of bedrock due to fractures

Weathering or alteration occurs along faults and fractures, developed irregularly in bedrock, by the action of underground water sometimes hot, producing extremely uneven physical properties in the rock near the surface. When one tries to assess the properties of these rocks in laboratory tests on collected samples, one can easily imagine that some properties will match those of the bedrock but others will not because of the presence of the discrete surfaces. Strictly speaking the differences between the properties of the samples and the bedrock itself will depend on the frequency of the fractures and their characteristics (whether clay is sandwiched in between or not, or whether the topography is undulating or not, etc.) and no general rules can be given. However, it is of prime importance to any engineering assessment of an area to ascertain the range within which such differences lie.

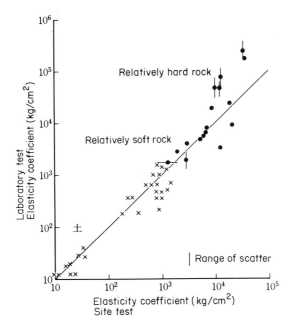

Figure 7.2   Changes in elasticity coefficients associated with consolidation of sediments (after Kojima, 1974*). × — borehole deformation test values and laboratory test values on cores under conditions corresponding to the depth from which they were collected. ● — on site loading test values, and laboratory test results for bedrock samples

The *elasticity coefficient* of sandstones and mudstones from all parts of Japan, shown in Figure 7.2, can be used as an illustration of these problems. The mechanical properties of rock slices were investigated using laboratory experiments on samples obtained from drill cores or in adits, and those of the bedrock itself by site tests such as *jack* or borehole *tests*. The more these rocks are consolidated, the greater is the influence of fractures, and so it is clear that considerable differences will arise between values for small samples of rocks and those for the bedrock. Even allowing for differences of test methods, in extreme cases up to a tenfold difference in the elasticity coefficients may be found. Such relationships are also encountered in shear strength, and the greater the variation in the strength of the rock itself and of the fracture plane, the greater will be the difference between core tests and site tests. In either case this tendency is stronger in relatively hard rocks.

However, this relationship must be examined for each individual discrete surface because the whole picture does not emerge solely by the above results. If a test is performed in which a force is applied perpendicular to the separation plane so that the rock on both sides is made to move in opposite directions relative to each other, the amount of displacement and the shear stress will show a relationship peculiar to each fracture. If an infill of clay, etc. follows the plane of the fracture, the two rock bodies will show a gentle stress–displacement curve with relatively small shear stress such as shown in Figure 7.3. However, in so-called rough fracture surfaces without such infill, shear stress increases as displacement progresses to maximum value. After passing this value, i.e. the peak shear strength, the shear stress drops suddenly and subsequently remains at an almost constant value even though displacement proceeds. The shear stress at this point is called the *residual shear strength*. The magnitude of this shear strength is also governed by the size of the normal stress applied perpendicular to the discontinuous plane. This relationship is shown in Figure 7.4. In other words the results of shear

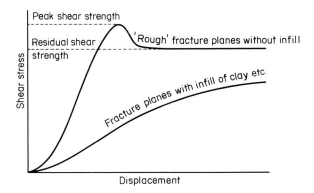

Figure 7.3   Relationship of displacement and shear stress under constant normal stress

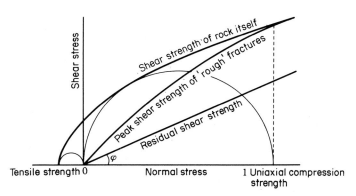

Figure 7.4   Relationship between normal stress and shear strength of 'rough' fracture planes without infill (after Ladanyi and Archambault, 1970*, etc)

tests demonstrate, in the shear stress–normal stress diagram, that there is a considerable difference between the shear strength of the rock itself, which does not incorporate fractures, and the peak shear strength of 'rough' surfaces together with the residual shear strength. At the same time we can also deduce that the larger the normal stress, the less the effects of the fractures, and if it is of a magnitude corresponding to the uniaxial compressive strength of the rock itself, the effects of fractures will be virtually eliminated. However, a fracture once subjected to a stress reaching the peak shear strength, will subsequently continue to be displaced with shear stresses equivalent to the residual shear strength, and the strength of the rock containing the actual fracture plane will finally be much less than that of the rock itself. Where there is an infill of clay, etc. it scarely needs to be said that the strength will decline further, but the extent of that decline will be governed by the clay content of the infill material, and types of clay minerals present. For example, by varying the normal stress and obtaining the *angle of internal friction* ψ relative to the residual strength in Figure 7.4, we are able to illustrate in Figure 7.5 an example of the relationship between tan ψ and the

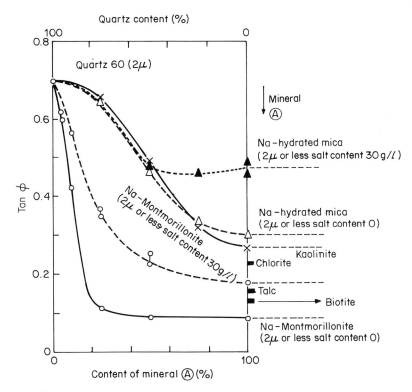

Figure 7.5 Relationship between the content of clay-sized mineral in fracture planes and the angle of friction relative to residual strength (Yoshinaka, 1977*). μ — microns

contents of various types of clay minerals. The higher the clay content, the lower tan ψ, and we can see that the extent of the decline varies considerably depending on the type of clay mineral. If the joints, faults or schistosity planes are orientated in one dominant direction, the behaviour of the rock containing them will probably vary remarkably depending on the direction of the externally applied force. An example of an experiment performed to assess the effects of such discontinuous plane orientation is shown in Figure 7.6. Samples were first prepared by consolidating a mixture of diatomite and plaster of paris. Fractures were made artificially in this and the simple shear tests shown in Figure 7.6(a) were performed. Ten fractures in a direction forming angle θ with the shear force were made and experiments were performed in which this angle and the vertical load were varied. The results are expressed using polar co-ordinates to yield Figure 7.6(b). The shear strength in rock samples without fractures was almost 2490 kg with a vertical load of 452 kg and is expressed by the semicircle in the diagram. On the other

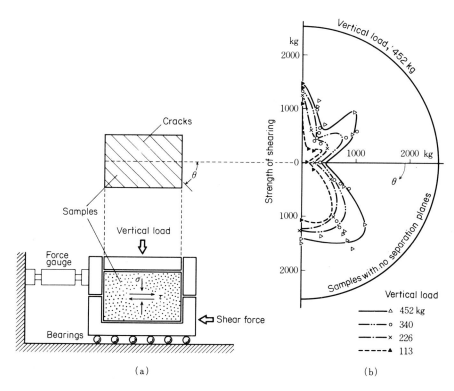

Figure 7.6   Relationship of shear strength and direction of fractures in shear tests (after Hayashi, 1965*). τ — shear stress. σ — normal stress

hand, the shear strength declined in samples with fractures, and anisotropy of the strength due to its direction was also remarkable. Mechanical anisotropy produced because the bedrock possesses discrete surfaces in various directions like this, is frequently encountered in site tests (Doboku Gakkai, 1975*, etc.).

To what extent will bedrock as a whole be mechanically irregular due to differences in strength and elasticity coefficients resulting from the presence of discrete surfaces, or due to mechanical anisotropy governed by the orientation of such surfaces? Let us consider an example of this, taking the elasticity coefficient as the parameter expressing the physical properties of the bedrock. As shown in Figure 7.7 there can in practice be divergences in the elasticity coefficient of up to 1000–fold even in granite which, of all the rocks found in the crust is thought to be relatively uniform, or in rock masses found in the relatively small areas over which engineering work is carried out. The bedrock can be classified into ranks A–F according to its competence by

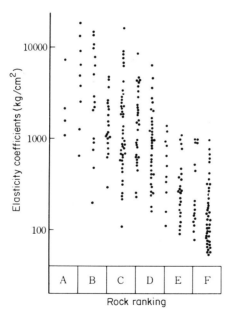

Figure 7.7 Scatter of elasticity coefficients in bedrock (Takahashi *et al.*, 1973*). Vertical axis — elasticity coefficient of bedrock measured in borehole. Horizontal axis — A–F — Bedrock rank based on observation of core sample. A — fresh hard granite with few cracks, declining in strength to B and C, and finally to F, extremely weathered granite

observing core samples collected from the site. (We will give a detailed discussion of rock classification of the abscissa described here in Section 7.4a.) Even if at first glance the cracks seem to have developed to the same extent, in practice there can still be a ten-fold variation in the elasticity coefficients, as can be seen from the scatter of values for each rank of bedrock. We must always remember that the degree of 'uniformity' will always incorporate at least this amount of scatter.

Discrete surfaces found in bedrock range in 'size' from localized or even latent structures to fault zones. The results of field measurements on a large scale may be regarded as the properties of the whole bedrock, which contains discrete surface of various sizes, and as we said earlier, they will not be equated with results obtained from samples tested in the laboratory. Such differences in the properties between small samples and large scale bedrocks have been termed the '*size effect*'. This is demonstrated by the fact that the larger the mass under consideration, the more numerous and diverse the discrete surfaces it contains and the greater the decline in its strength and elasticity coefficient. Alternatively, one can imagine that the non-uniformity due to small planes of discontinuity is averaged out with the whole mass, and the bedrock as a whole seems to be apparently more uniform and the scatter of measured values declines.

Very little data on size effects are available, and many points remain unclear. For example, in Figure 7.8 which illustrates this effect on strength we

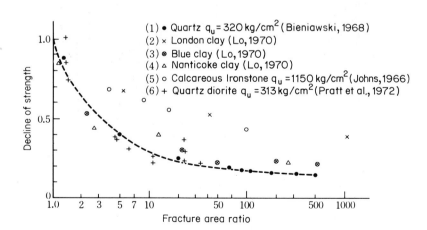

Figure 7.8   Size effect on strength (Yoshinaka, 1976*). The fracture area ratio shows the ratio of the shear area of ordinary site test to the shear area of laboratory test. Clay in this diagram may for practical purposes be regarded as mudstone containing cracks. The curve approximates a relationship between the two with an exponential function and is an example applicable to material (1). qu — uniaxial compressive strength

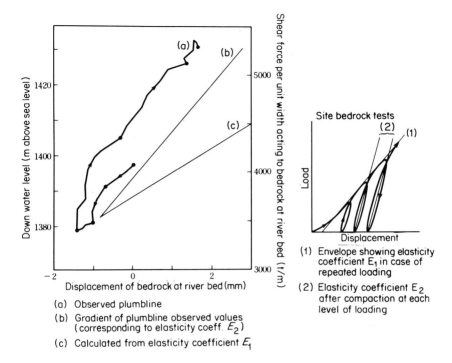

(a) Observed plumbline
(b) Gradient of plumbline observed values
    (corresponding to elasticity coeff. E_2)
(c) Calculated from elasticity coefficient E_1

(1) Envelope showing elasticity coefficient E_1 in case of repeated loading
(2) Elasticity coefficient E_2 after compaction at each level of loading

Figure 7.9   Bedrock deformation at site of Kurobe River No. 4 Dam (Doboku Gakkai, 1975*)

find that, for the same type of rock, the greater the dimensions the more the strength falls over sizes ranging from a core sample to the area of the site test.

Let us now consider deformation of the bedrock in a somewhat broader concrete example. Figure 7.9 is an example of test and observed results at the site of a dam, and shows records of a plumb line† provided with the intention of ascertaining the deformation behaviour of the dam wall and the surrounding bedrock for the purposes of safety management after construction. One may use it to assess what sort of deformation the increased load caused by the build-up of water behind the dam, will actually produce in the bedrock for about 100 m around the dam wall. In this instance the elasticity coefficient of the bedrock was obtained on the basis of a site test on a scale 1 m in diameter during surveys at the design stage of the dam wall. The elasticity coefficient obtained from displacement of the plumb line corresponded very well with that obtained in bedrock tests, when a load very similar to the increasing load of the dam wall construction, i.e. loads corresponding to the rise and fall of

† A hole is made in the bedrock and a weight attached to steel wire hangs in it. Relative displacement in the bedrock over a specified area is measured by its movement.

the water level were applied repeatedly (E_2 in Figure 7.9) to the bedrock. In the range of this example, differences due to the dimensions of the two were seen to be quite small.

The coherence of the bedrock is something else which must not be forgotten when considering the mechanical properties of surface rock formations. For example, a lack of coherence is a problem of overriding importance, especially when discussing the stability of slopes or considering how to go about excavating tunnels. Various survey methods have been devised to assess this. Because one must examine the properties of coherent bedrock in bedrock tests, an operation is performed at the time of the test to completely remove any loose pieces and to compact the sample by repeated loading. Degrees of coherence are infinitely variable, and one cannot speak generally of how great or extensive it is. Let us consider an interpretation of test results such as those given in Figure 7.10. This diagram compares actual with theoretical values obtained on the assumption that the bedrock is an elastic body, for deformation directly below the centre line of a diaphragm in bedrock tests (jack test) on diorite undertaken in a survey shaft near the surface. When one compares the increases in strain with loading at three levels and repeated loading, the increase in strain at depths of less than 40–50 cm below the surface was seen to fall when the load increased. If the rock was an elastic body, the increase in strain should be uniform. In other words, because this rock body is incoherent near the surface it does not behave like an elastic body. However, when a repeated force was applied, the spaces between cracks were compacted, and the rock as a whole became more like an elastic body. On the other hand, when the bedrock is of good quality with relatively few fractures, it behaves like an elastic body at greater depths; and observed strain increases are consistent to those corresponding to the elasticity coefficients measured in core samples from there. The properties of rocks containing such fractures are extremely complex as they vary depending upon the stress operated and pressure due to loading as well as the degree of looseness. Generally speaking if fractures are present, the properties of a given bedrock as an elastic body are weakened. As the load increases, the rock will be compact, the incoherence will be reduced and the more it will manifest the characteristics of an elastic body. Consequently as in this example, one will then observe a tendency for the properties of bedrock with no looseness to resemble more closely those of the core samples.

7.4   Methods of evaluating bedrock

The strength and elasticity coefficients of bedrock decline depending on the size and frequency of discrete surfaces, which, in practice, must be assessed as a part of an overall picture of the properties of the rock concerned. When considering discrete surfaces in the rock, one must also take into account the

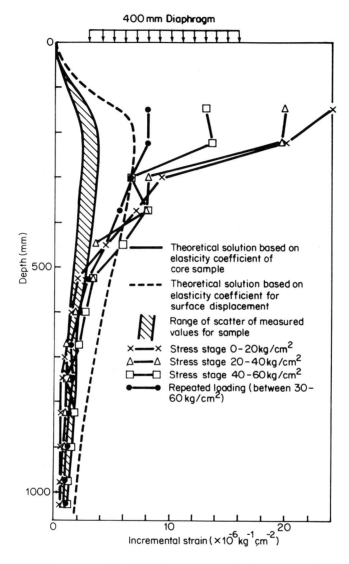

Figure 7.10 Displacement directly below centre of diaphragm (after Iida and Kobayashi, 1973*)

size of the rocks or the bedrock appropriate to the dam, tunnel or cutting. Let us consider Figure 7.11. In this diagram *a*, *b* and *c* are three types of discrete surfaces in the geologic body on different scales. One can see differences in their direction, size and density. Relative to the scale of the dam wall the

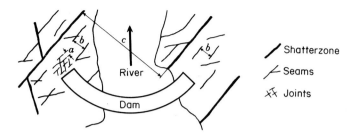

Figure 7.11   Grouping of discrete surfaces in geological body of various thicknesses in
the bedrock (Doboku Gakkai, 1977*)

group of joints *a*, which have the smallest size, can be ignored, because the
effects of such fractures are assessed directly by site bedrock tests as the
averaged properties of the whole of the bedrock including these joints. This
kind of test cannot be used, however, to evaluate the effect of seams and
faults on the larger scale like *b*. It would be too complicated to analyse their
properties and orientation individually. For groups of discrete surfaces in the
geologic body on this sort of scale we have to prepare simplified models and
discuss their effects on the basis of model experiments or mathematical
analysis using the finite element method which we discuss later. On the other
hand, we will be able to examine directly the essential mechanical properties
of the fault shatter zones and the seams by the site tests. Measurements thus
taken directly in site tests furnish the basis for an assessment of the bedrock.
But actually there are limits as to cost and test locations, and the distribution
of mechanical properties in the rocks under consideration as a whole is
difficult to understand. What we can do, therefore, is to look at outcrops,
survey adits and drilled samples, classify the quality of the bedrock, and
devise methods of comparing the observed results with those obtained by a
few site tests. This form of bedrock classification plays a vital role in
ascertaining the distribution of mechanical properties in the area under
consideration. In practice physical values such as the elastic wave velocity
usually measured at the primary stage of the survey are incorporated with
these so that an overall picture of the bedrock can be built up.

(a)   Rock classification

The rock classification is essentially based on factors which reduce the
strength or elasticity coefficient of the bedrock (deterioration factors). After
examining the factors on a site, these are then brought together and the
results are expressed as objectively as possible so as to correlate them with the
physical properties of the bedrock under consideration. However, the

emphasis of rock classification varies with the type of project envisaged, and the methods of the classification proposed are also various; specifically dependent on whether dams, tunnels or cuttings are involved. One example is shown in Table 7.2. In this case, we find the factors which bring about deterioration and classify them into many levels according to their effectiveness. These are grouped and the results are applied to an area in order to give a grade and general assessment of the bedrock. There is an overwhelming

Table 7.2 Bedrock classification and assessment by combination methods (after Okamoto and Akie, 1966*)

(a) Bedrock classification standards (for andesite)

Factor	Rank	Details
Hardness of rock slice	A	Hard (1)
	B	Some hard, some soft, the whole slightly soft (2)
	C	Soft (3)
Spacing of cracks	I	More than 50 cm
	II	50 – 15 cm
	III	Less than 15 cm
Condition of cracks	a	Tightly compact
	b	Open
	c	Clay infill

(1) Sparks when struck with hammer
(2) Can be broken with one hard blow of hammer
(3) Will shatter with hammer.

(b) Bedrock assessment (Shimouku dam)

Assessed rank	Evaluation	Combinations of ranks
A	Good	AIa, AIb, BIa, BIb
B	Fair	AIc, AIIa, AIIb, BIc, BIIa, BIIb, CIa
C	Rather poor	AIIc, CIb, CIc, CIIa, CIIb
D	Poor	Other combinations

emphasis on factors such as the hardness of samples, the degree of weathering, the spacing of fractures and the nature of fracture planes. Classifications based on the rock type and seismic wave velocity, etc. are also commonly used. When undertaking a classification, the scope of the relevant categories and the criteria of assessment are governed in various ways but particularly by the proposed works. There has been a tendency in recent methods of rock classification to try and award marks corresponding to the level of each factor, and to assess the rock on the basis of the sum or product of the scores thus obtained. There are also several instances in which particular consideration is given to the properties of fractures as relevant factors. For example, Barton's (1974*) method takes scores for the level of each factor shown in Table 7.3 and calculates the assessment value Q as

$$Q = (RQD/Jn) \times (Jr/Ja) \times (Jw/SRF) \tag{7.1}$$

RQD: Rock quality designation. In a boring core, 1 m long, proportion of total length of core rods 10 cm or longer to the whole. It represents the interval of cracks in the core rod, if the bedrock is hard (see Table 7.3). SRF: Stress reduction factor, see Table 7.3.

Table 7.3   Bedrock classification by points system. Factors and scores to be evaluated in classifying rock rank (simplified from Barton *et al.*, 1974*) (see equation 7.1)

Factor	Summary of ranking and assessment
RQD	10 when 10% or less; and one uses the value of RQD
Jn (no. of sets of discrete surfaces)	Scoring from no fractures, 1 set, 4 or more sets shatter zones into 9 divisions, Scores 0.5–20.0
Jr (roughness of discrete surfaces)	Points awarded (4.0–0.5) for fine classification into conditions such as no continuous cracks, wavy, platy, clay infill etc. However, score 1.0, if spacing of fractures is greater than 3 m.
Ja (infill or alteration of fracture plane)	Two major divisions in infill present or not. If not, score 0.75–4.0 by assessing degree of alteration of the plane. If infill is present, score 4.0–20.0 according to type, thickness and stratification of infill.
Jw (reduction by interstitial water)	Score 1.0–0.05 in 6 divisions from 'dry' to 'much inflow'.
SRF	If weak layers present, depending on case as to how they cut the excavation plane, and if not present classifying into hard rock, plastic rock or swelling rock, score 0.5–20.0.

He suggested that as a criterion of assessment for earth pressure and support work in tunnels (e.g. frameworks supporting shaft walls, etc.) the value of Q ranges from 0.001 for degenerated bedrock with marked tendency to swelling, to 1000 for good quality rock with no fractures. The first term in equation (7.1) expresses the average dimensions of the bedrock block, the second term the characteristics of the friction between blocks and the third one the operational stress. The more the properties of rock are expressed quantitatively by such method, or the more emphasis we place on the nature of fracture planes, the longer it takes to examine and record the rock. Many methods are therefore impractical for reasons of the feasibility of survey and the time involved. In practice, however, simpler methods as proposed by Tanaka (1974) are widely used, in which the levels of geological factors such as the degree of weathering and hardness are brought together.

Figure 7.12 shows the elasticity coefficient of a bedrock formed from sandstone and mudstone, and compares the value obtained from a correlation between the ranking based on rock classification of core samples and the score method of assessment with the values actually measured by shaft deformation tests at the place where the core samples were collected. The bedrock ranking is an ordinal scale and begins with group A which is the

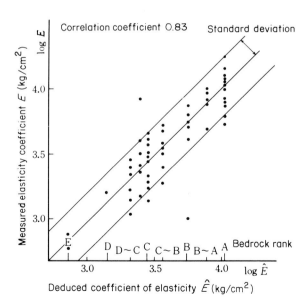

Figure 7.12   Relationship between elasticity coefficients observed and deduced from correlation analysis of rock classification (Kojima, 1975*)

freshest and least fractured rock and continues in steps of increasing qualitative 'poorness'. However, looking at this conveniently calibrated system with an equal interval, one can see that it maintains an almost linear relationship with the logarithm of the elasticity coefficient. A similar relationship can be discerned in the results of block shear test in bedrock adits shown in Figure 7.13. Considering the test values for rocks of each identical rank, we obtain a straight line expressed by $\tau = C + \sigma \tan \psi$, i.e. the relationship between the shear stress τ and the normal stress σ as in the diagram. Thus a close correlation exists between the rock ranking and the physical properties, though there is a certain scatter of values. If one further imagines that a fairly considerable scatter will also be observed even in 'numerical bedrock evaluations' based on the results from actual shaft tests, the evaluations of physical values based on rock ranking may be said to be sufficiently practical.

The distribution of mechanical properties of rocks containing fractures may be determined from a fairly limited number of bedrock test values by using a rock ranking system in places where the bedrock can physically be seen with survey adits, boreholes or surface outcrops. These results can also be used as a basis for assessing areas, which cannot be seen, by incorporating the results of seismic survey, and they can then all be drawn together. Figure 7.14 shows

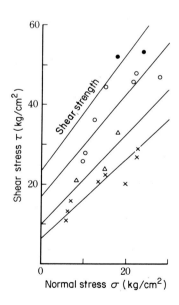

Figure 7.13   Example of measured shear strength for bedrock ranking B–E. (Iida, 1978*). Ranks ● B, ○ C, △ D, × E

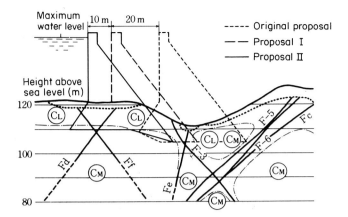

Figure 7.14 Geological profile showing bedrock rank distribution around dam site
(Noda *et al.*, 1976*)

an example of a geological engineering profile of the river bed section of a
proposed dam site (consisting mainly of green tuff) obtained in this manner.
Table 7.4 illustrates the values for each property corresponding to the rock
ranks.

(b) Bedrock models

It is important to devise models of the bedrock appropriate to the objective
and to analyse its strength and potential for deformation, if we are going to
predict the fractures and the deformation which could be caused in the
bedrock by excavating or by placing structures on top of it, and moreover to

Table 7.4 Physical properties for each sector used when bedrock deformation
associated with dam construction and safety *vis-à-vis* collapse were calculated on the
basis of a model (see Figure 7.14, after Iijima, 1975*)

Geology	Bedrock rank	Elasticity coefficient (t/m^2)	Shear strength (t/m^2)	Friction coefficient	Notes (Figure 7.14)
Good bedrock	Cm	20 000	90	1.0	
Alteration part	Cl–Cm	10 000	60	0.8	
Faults	D	2 000	50	0.6	F–3, F–5
Concrete		200 000			

plan and execute the work as safely and economically as possible. As we discussed above, it is usual to analyse the bedrock with rock ranking and subsequently to substitute the bedrock by a continuous averaged-out body. A growing trend in recent years, however, has been to try to devise models directly of the fracture systems themselves and to analyse these (Goodman, 1976*, etc.).

Figure 7.15(a) shows the rock ranking profile shown in Figure 7.14 and attempts to devise a model of it so that mathematical analysis can be performed using the finite element method. The existence of discrete surfaces

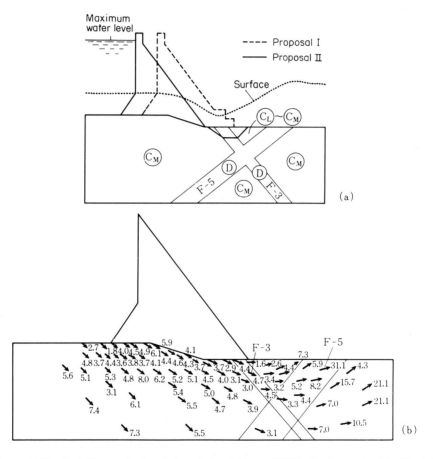

Figure 7.15   Stability analysis of dam (after Iijima, 1975*; Noda *et al.*, 1976*). (a) Model of bedrock based on Figure 7.14. (b) Calculated safety factor for each point towards shear failure. Arrows – direction of slippage. Figures – safety factor (force resisting slippage/force encouraging slippage)

in the bedrock of a few metres in scale, such as joints, was previously incorporated into the rock rank and the whole was regarded as a continuous body. Thus the practical problems with the discontinuous planes relate to faults on the largest scales. Because there are limits on the number of elements when devising models for finite element methods, the geological structures must be simplified or exaggerated fairly drastically. There is also a considerable range in the properties corresponding to any particular rock ranking. It is therefore meaningless to allow consideration of too detailed structures. For example, the width of faults F–3 and F–5 in Figure 7.14 is greatly exaggerated in Figure 7.15, whereas Fd, Fe and Ff were of such small scale that they were judged unlikely to have any major safety implications for the dam, and so were left out. In the model the discontinuity of F5 is expressed by a 10 m wide fracture zone containing F6 and Fc, and is treated by subdividing it into groups of elements with low elasticity coefficients. If F3 with its 1 m wide fracture zone was treated in the same way, the number of elements would be excessive, and so its width is exaggerated three-fold. To compensate for this, care is taken to ensure that the elasticity coefficient of the fracture zone is magnified 2.5 times so that the total deformation of the fault zone as a whole remains the same. When actually constructing the dam, CL on the surface in Figure 7.14 lies above the line of excavation and this section will be removed. Ultimately the bedrock under consideration comes down to rank Cm to the left of fault F–3 and rank Cl–Cm to the right. The mechanical properties needed for an analysis of each one were generally obtained from the rock tests and are given in Table 7.4. When the above model is analysed, displacement in each element, the principal stress distribution and the safety level of each element relative to shear fracturing are obtained. An example of results calculated in the early design stage is shown in Figure 7.15(b). The arrows indicate the directions of slippage and the figures express the safety level in respect to slip fracture. (However these differ from the average safety factor described in Section 7.1.)

(c)   The stability of slopes

When excavating the bedrock and making a huge slope, whether or not that slope is stable or how it can be stabilized are problems which we frequently encounter as the project design progresses. If the bedrock contains fractures, the following type of vector analysis is occasionally used in stability analyses focusing on specific dominant groups of fractures. Figure 7.16 illustrates the easiest case representing slip phenomena along a fracture plane. If the slope of the fracture plane is taken to be θ, the shear resistance angle to be ψ, the weight of a block on the fracture plane to be W and the safety factor to be F_s then clearly, from the diagram

$$F_s = \frac{\text{force resisting slippage}}{\text{force encouraging slippage}} = \frac{S}{Df} = \frac{C + W \cos \theta \cdot \tan \varphi}{W \sin \theta} \tag{7.2}$$

If cohesion C of the fracture plane is small, it may be ignored; and the condition for no slippage from equation (7.2) gives us:

$$F_s = \frac{\cos \theta \cdot \tan \varphi}{\sin \theta} = \frac{\tan \varphi}{\tan \theta} \geq 1, \qquad \text{therefore } \theta \geq \varphi$$
$$\tag{7.3}$$

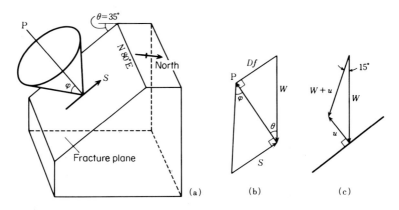

Figure 7.16   Slippage along fracture planes and forces at work there (Attewell and Farmer, 1976*). (a) Friction cone. (b) Forces acting on fracture plane. (c) Effects of interstitial water pressure in acting on fracture plane (in this case, direction of $W + u$, and W is 15°)

It will be stable if the inclination θ of the fracture plane is less than the shear resistance angle ψ. *The friction cone* in the diagram illustrates the area which satisfies this and is expressed as in Figure 7.17 on a stereonet. In other words, it is stable when an acting force W shown in Figure 7.16(b) exists in the small circle in the diagram. On the other hand, if water is present, an interstitial water pressure u will act as shown in Figure 7.16(c) on the rocks over this slope. As a result, the direction of the force applied over the slope will shift slightly, and the direction will emerge on the outside of the small circle expressing the stable region, as shown in Figure 7.17. In short, one may conclude that this slope will become unstable due to the effects of interstitial water pressure. In either case, however, we assume cohesion of the fracture plane to be small, fracturing inside the bedrock block to be absent, and

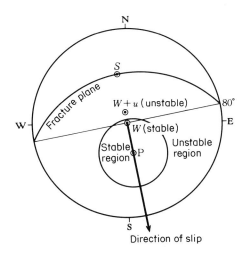

Figure 7.17   Stability analysis relating to slippage along fracture plane. Angle of shear resistance of fracture plane 40°, projected on the upper-hemisphere of the Wulf's net. See Figure 7.16

slippage to occur in the direction of maximum slope. There are very few instances in the bedrock of Japan where all the above assumptions apply, but it is relatively easy to expand the method for groups of fractures in several directions and to undertake a three-dimensional treatment. It therefore finds many applications in the practical consideration of slope stability analysis (Doshitsu Kogakkai, 1974*; Attewell and Farmer, 1976*; Goodman, 1976*).

Keiji Kojima

References

Attewell, P. B. and Farmer, I. W. (1976): *Principles of Engineering Geology*, 1045 pp. Chapman & Hall, London.

Barton, N., Lien, R. and Lunde, J. (1974): Engineering classification of rockmass for design of tunnel support. *Rock Mechanics*, **6**, 189–236.

Doboku Gakkai (Civil Engineering Assoc.) ed. (1975): *Rock Mechanics for Civil Engineers*, 676 pp. Doboku Gakkai.

Doboku Gakkai (1977): *Geological Surveys for Dams*, 188 pp. Doboku Gakkai.

Doshitsu Kogakkai ed. (Soil Mechanics Assoc.) (1974): *The Mechanical Properties of Rocks and their Application to Design and Execution*, 838 pp. Doshitsu Kogakkai.

Goodman, R. E. (1976): *Methods of Geological Engineering in Discontinuous Rocks*, 472 pp. West Publishing Co., New York.

Hayashi, M. (1965): Shear resistance, anisotropy and dilatancy of laminated bodies, and mechanisms thereof. *Proceedings of Third Symposium on Bedrock Mechanics*. Doboku Gakkai, 17–21.

Iida, R. and Kobayashi, S. (1973): Relationship of bedrock looseness and behaviour in site loading tests. *Summary of Proceedings of 8th Symposium on Bedrock Mechanics*. Doboku Gakkai, 36–41.

Iida, R. (1978): *Bedrock Mechanics in Civil Engineering*, 318 pp. Shokokusha, Tokyo.

Iijima, H. (1975): Geological models of dams and FEM analysis. Oyo Chistitsugakkai Kenkyu Happyokai (Abstract in Applied Geology Assoc.), 31–32.

Kojima, K. (1974): Problems of survey methods for soft rock. *Tsuchi to Kiso*, **22**, 6, 53–8.

Kojima, K. (1975): Distribution of bedrock properties and the human eye. *Seko Gijutsu*, **8**, 5, 52–4.

Ladanyi, B and Archambault, G. (1970): Simulation of shear behaviour of a jointed rock mass. *Proc. 11th Symp. on Rock Mechanics* (AIME), p. 105.

Noda, S. *et al.* (1976): Geological models for dams and geological engineering surveys by FEM. *Oyo Chishitsu*, **17**, 100–104.

Okamoto, R. and Akie, A. (1966): Attempts at bedrock classification at dam sites. *Doboku Gijutsu Shiryo*, **6**, 423.

Takahashi, K. *et al.* (1973): Assessing the mechanical properties of granites using borehole measurements. *Proceedings of 4th Japanese Symposium on Rock Mechanics*. Doboku Gakkai, 19–24.

Yoshinaka, R. (1976): Size effects on the strength of bedrock. *Seko Gijutsu*, **9**, 8, 58–60.

Yoshinaka, R. (1977): Friction characteristics of discrete surfaces in geologic body. *Seko Gijutsu*, **10** 5, 56–9.

Further Reading

General textbooks on engineering geology not dealing solely with the topics covered in this chapter include:

Tanaka, H. (1965): *An Introduction to Engineering Geology*, 169 pp. Sankaido.
Takahashi, H. and Miyazaki, S. (1970): *Engineering Geology*, 494 pp. Kyoritsu Shuppan.

A good textbook which covers related fields such as soil and bedrock mechanics and hydraulics together with methodology is that of Attewell, B. B. and Farmer, I. W. (1976*). On the subject of dam construction, to which engineering geology is most comprehensively applied, Zaruba, Q. and Mend, V. (1976): *Engineering Geology*, 504 pp. Elsevier, Amsterdam deals with examples which follow the line of thinking of the Doboku Gakkai (1977*) in the sequence of processes, etc.
A general geology textbook for technicians is:

Takahashi, H. (1974): *Geology for the Civil Engineer*, 226 pp. Kashima Shuppankai.

Books dealing with basic geological methods such as geologic mapping, stereographic projection and rock identification methods include:

Seki, Y. (1976): *Petrology for the Construction Technician*, 149 pp. Kyoritsu Shuppan.
Blyth, F. G. H. (1967): *A Geology for Engineering*. Edward Arnold, London.
Books on bedrock mechanics starting from elastic and plastic theory to explanation of the mechanical properties of bedrock include:

Yamaguchi, U. and Nishimatsu, V. (1977): *Introduction to Rock Mechanics*, 2nd ed., 266 pp. Todai Shuppankai.

Talobe, J. (1957): Translated by Shindo, K. (1967): *Bedrock Mechanics*, 411 pp. Morikita Shuppan.

Jaeger, J. C. and Cook N. G. W. (1969): *Fundamentals of Rock Mechanics*, 585 pp. Methuen, London.

The organization of bedrock mechanics is summarized in Doshitsu Kogakkai ed (1974*) and Doboku Gakkai ed. (1975*) which are also useful as references for the topics covered in this chapter. Similarly, Iida, R. (1978*) deals with methods and interpretation from bedrock tests to analysis, and Goodman, R. E. (1976*) with rock mechanics and the engineering aspects of fractures. These are recommended to the reader interested in the subject matter of this chapter.

Finite element analysis has come to be widely used not only in the problems dealt with in this chapter but in the analysis of the physical properties of many types of model. A good reference work is:

Zienkievicz, O. C. (1971): *Matrix Finite Element Methods in Basic Engineering*. 575 pp.

A general introduction to the subject is contained in:

Togawa, H. (1974): *Introduction of Finite Element Methods Based on Fortran*, 220 pp. Science Co.

Geological Structures
Edited by T. Uemura and S. Mizutani
©1984 John Wiley & Sons Ltd.

Appendix

Descriptive Terms for Geological Structures

Innumerable specialized terms are used in geology to describe structures accurately and to discuss the mechanisms by which they developed. These include all manner of words ranging from simple geometric descriptions to those indicative of processes or origins. In this appendix, partly due to restrictions on space, we explain briefly some of the particularly basic and well known terminology.

1. Elements of geological structures

The basic elements of which geological structures are composed are lines and planes, and hence give rise to, 'linear structures' and 'planar structures'. Both of these are measured and described on the basis of their geographical direction and horizontal plane, and are often expressed by plotting them on a stereographic net or equal-area net.

The form of a plane surface such as a bedding plane is expressed by its *strike* and *dip*, and is plotted as the trace of the plane surface itself or its pole on a stereographic net, etc. The attitude of a straight line is shown by the angle it makes with the horizontal (*plunge*) and the orientation in which it descends. Linear structures on a specific plane, for example, striations on a fault plane or a linear fold axis (see Figure A.10) may sometimes be expressed by the relationship of the two, using the angle, measured within the plane, that the linear structure makes with a horizontal line drawn on the plane. This angle is known as the *rake* or *pitch* (Figure A.1).

Various characteristic arrangements are encountered in describing planar and linear structures. For example, if the poles to planes tangential to the folded surface of a cylindrical fold (Figure A.9), are plotted on a stereographic net, they will fall on a single great circle, and the pole to the great circle will correspond to the fold axis. The diagram thus plotted, using poles to planes is known as a π diagram.

289

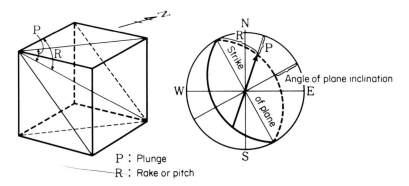

Figure A.1   Attitude of line and plane and stereographic projection thereof

In the case of real cylindrical folds, the poles to the folded surface do not appear on a single circle on the π diagram, but are commonly distributed over a certain range along the trace of the circle. Such a great circle-type of concentration pattern is known as a girdle.

If folded surfaces of a cylindrical fold are not expressed as their poles but as great circle traces on a stereographic net or equal-area net, each great circle will intersect at one point, which will indicate the direction of the fold axis. Plotting such traces of planar structures produces a β *diagram* (Figure A.2).

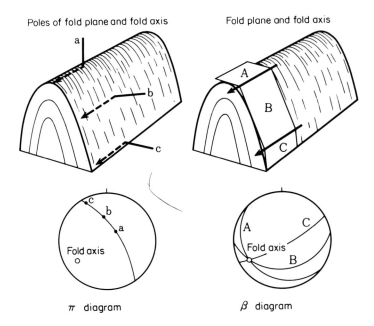

Figure A.2   π diagram and β diagram (lower hemisphere projection)

A pattern of distribution in which the plotted points are gathered around almost one point is called the 'maximum'. A point maximum expresses parallel linear structures or the poles of parallel planar structures.

Mutual relationships such as parallel, orthogonal, oblique or *en échelon* can be distinguished in the arrangement of linear structures on a plane surface. The *en échelon* arrangement may be further subdivided into the right-handed (italic-m) type and the left-handed (back-hand-writing-m) type as illustrated in Figure A.3.

2. Faults

The terminology and classification of faults is based fundamentally on the characteristics of fault movement obtained by observation.

If one classifies faults according to the relative movement of the geologic bodies on either side of the fault plane, the fault can be divided into either rotational or translational, and may be further distinguished on the basis of the direction of the movement as in Figure A.4. These terms are applied when the amount of actual displacement of the geologic bodies on either side of the fault plane, i.e. the slip, can be ascertained.

If striation is left on the fault plane, then the direction of relative movement along the fault plane can be assessed, but it is not possible to determine the movement vector of the geologic bodies on either side from this alone. Minute steps almost at right angles to the direction of relative movement may

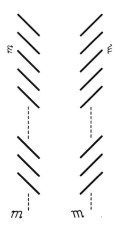

Figure A.3 *En échelon* arrangements of right-handed (italic-m) type and left-handed (back-hand-writing-m) type. Linear structures shown by the short lines are parallel to each other. The direction of the row, i.e. the straight line connecting the centre point of each constituent strand intersects the constituent strand obliquely

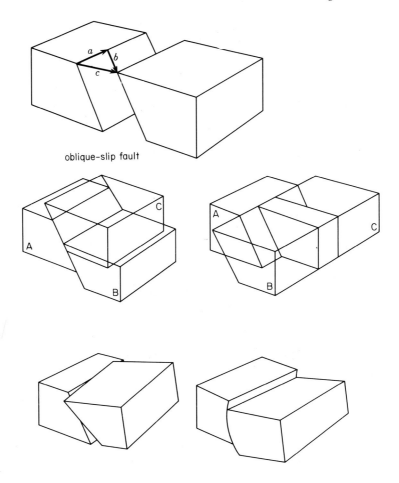

oblique-slip fault

Figure A.4   Classification of fault based on the direction of relative displacement
(slip)

(a) component of displacement in direction of strike of fault plane (strike slip)
(b) component of displacement in direction of dip of fault plane (dip slip)
(c) amount and direction of actual displacement (net slip)

Dip slip fault   Strike slip fault
Gravity or normal fault: A–B   Right-handed strike slip fault: A–B
Thrust or reverse fault: A–C   Left-handed strike slip fault: A–C

Rotational faults: hinge fault (axis of rotation perpendicular to fault plane) and
cylindrical fault (axis of rotation parallel to fault plane)

be sometimes produced on the fault plane, and they are believed to face in the direction of movement of the opposite block. However, such as assessment is not invariably correct. It would seem to vary according to the formative processes and the mechanism of the fault.

If the direction of relative movement is unclear, the fault may be classified on the basis of its *separation*, i.e. the apparent displacement of characteristic strata and rocks (Figure A.5).

In most faults it is virtually impossible for us to assess the actual movements which result in such relative and apparent displacement, and there is very little terminology to cover this problem. However, in some cases of large-

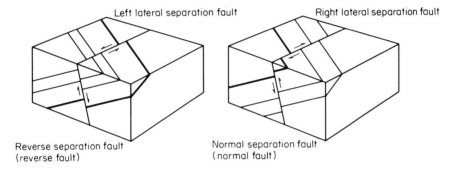

Figure A.5   Classification of fault based on apparent displacement (separation)

scale low angle overthrusts it is believed that the overlying block on the fault plane perhaps slides up unidirectionally whereas the lower block hardly moves at all.

Faults may further be classified into strike fault, dip fault and oblique fault based on the attitude of the fault relative to the attitude of the adjacent strata, or alternatively are sometimes divided into longitudinal fault, transverse fault, and diagonal fault from their relationship to the regional geological structures.

3.   Folds

Folds are planar structures formed when flat layered structures such as bedding planes are deformed. If one looks at a single layered fold profile, it usually possesses a section or, to be exact, a point at which the curvature is greatest. The line connecting these points, or hinges, is called the hinge line or fold axis. The wings, flanks or limbs lie to either side of this where the

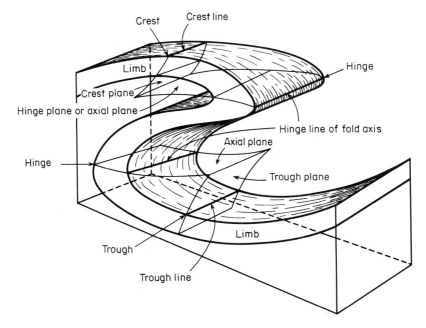

Figure A.6   Terminology for each part of a fold

curvature is much less. The point of no curvature near the mid-point of the limb is called the *inflection point* and the plane which contains all the hinge lines for each fold plane is the *axial plane* or axial surface. The hinge line and the axial plane form the most basic geometrical elements to the folded structure (Figure A.6).

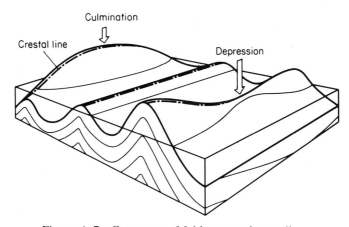

Figure A.7   Curvature of fold axes and crest lines

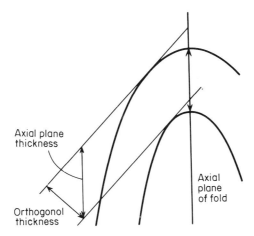

Figure A.8   The thickness of a folded layer (after Ramsay, 1967*)

Another group of terms is used for each section of a multi-layered fold seen in its vertical profile. Connecting the *crests*, which are at the highest position above a horizontal standard line, and the troughs at the lowest position gives the definition of the crest line and crest plane, and trough line and trough plane, respectively. These are particularly important to describing the geological structures in oilfield regions. Fold axes or hinges or crest lines are not generally straight lines but are themselves curved. When a crest line is convex in the upwards direction this gives rise to a *culmination*, whereas the relatively sunken part forms a *depression* (Figure A.7).

To illustrate folded structures on geological maps, instead of the fold axis or the crest line it is common to use an *axial trace* or a *crest trace* which is the line of intersection of the axial plane or crest plane with the earth's surface plane.

To illustrate the characteristics of thickness and form at each position of the fold for single-layered folded structures a *tectonic profile* is drawn at right angles to the fold axis. In such profiles the thickness of each folded layer is shown as the distance between parallel tangents drawn on the upper and the lower boundary of each layer, i.e. the *orthogonal thickness*, and as the *thickness* measured *parallel to the axial plane* (Figure A.8).

Morphological classification of folds may be based on various criteria, and the most basic approach looks at the properties of geometrical elements. Figure A.9 demonstrates the classification based on the geometrical properties of the hinge line and the axial plane. Horizontal plane cylindrical folds from amongst these may be further classified as in Figure A.10 using criteria

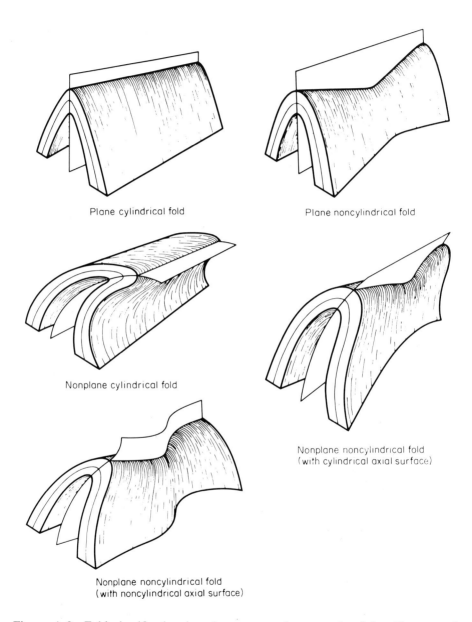

Plane cylindrical fold

Plane noncylindrical fold

Nonplane cylindrical fold

Nonplane noncylindrical fold
(with cylindrical axial surface)

Nonplane noncylindrical fold
(with noncylindrical axial surface)

Figure A.9   Fold classification based on geometric properties (after Turner and Weiss, 1963*)

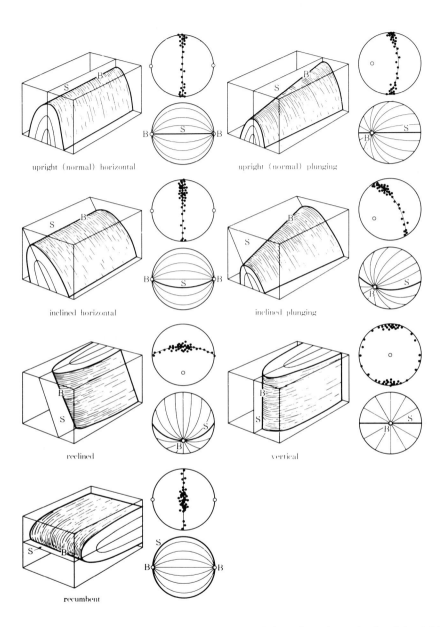

upright (normal) horizontal

upright (normal) plunging

inclined horizontal

inclined plunging

reclined

vertical

recumbent

Figure A.10   Classification of plane cylindrical folds based on the attitude of the fold axis and axial plane (terminology after Fleuty, 1964*). B: Fold axis or hinge line. S: Axial plane

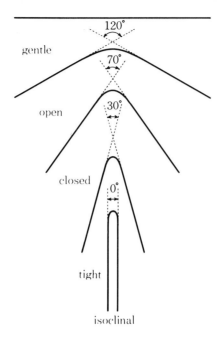

Figure A.11   Classification of folds by interlimb angle (terminology after Fleuty, 1964*)

based on the attitude of the rectilinear hinge line and planar axial surface relative to the horizontal plane. A further type of classification is possible based on the interlimb angle, as shown in Figure A.11 and parallel or *similar folds* (Figure A.12) may be distinguished from the manner in which the layer thickness changes from the hinge towards the limbs (see Figure 5.10).

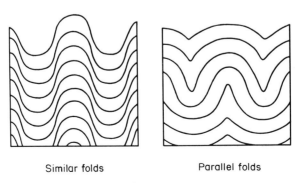

Similar folds   Parallel folds

A.12   Similar folds and parallel folds

No indications of the scale are given in any of the diagrams in this section. The reason for this is that the features described can be found in geological structures of various scales, and that the terminology is normally used irrespective of their size. However, it is the accepted practice that certain terms such as low angle overthrust, *recumbent fold*, culmination or depression are used only for large scale structures such as are found, for example, in orogenic zones.

Shinjiro Mizutani and Takeshi Uemura

References

Billings, M. P. (1972): *Structural Geology*, 3rd Ed., 606 pp. Prentice-Hall, Englewood Cliffs, New Jersey.

Dennis, J. G. (1967): *International Tectonic Dictionary. English Terminology*, 196 pp., Am. Assoc. Petrol. Geol., Mem., 7, Tulsa, Oklahoma.

Dennis, J. G. (1972): *Structural Geology*, 532 pp., Ronald Press, New York.

Fleuty, M. J. (1964): The description of folds, *Geol. Assoc. (London) Proc.*, **75**, 461–92.

Heibonsha (1970): *Dictionary of Geology*. 1540 pp. Tokyo.

Hobbs, B. E., Means, W. D. and Williams, P. F. (1976): *An Outline of Structural Geology*, 571 pp., John Wiley, New York.

Kokinshoin (1973): *New Dictionary of Geology*. Vol. III, 799 pp. Tokyo.

Ragan, D. M. (1973): *Structural Geology: An Introduction to Geometrical Techniques*, 2nd Ed., 208 pp., John Wiley, New York.

Ramsay, J. G. (1967): *Folding and Fracturing of Rocks*, 568 pp., McGraw-Hill, New York.

Turner, F. J. and Weiss, L. E. (1963): *Structural Analysis of Metamorphic Tectonites*, 545 pp., McGraw-Hill, New York.

Whitten, E. H. T. (1966): *Structural Geology of Folded Rocks*, 678 pp., Rand McNally, Chicago.

Index